David Bohm's Implicate Order:

New Frontiers in the Evolution of Consciousness

Shelli Renée Joyce

B.S. Electrical Engineering
M.A. Indian Philosophy
Ph.D. Philosophy and Religion

Published by the Viola Institute
Viola, California

This book is dedicated to my many mentors including Dr. John Lilly, Alan Watts, Dr. Haridas Chaudhuri, and Dr. Brian Swimme, Dr. Allan Combs, and Dr. Dean Radin. However I would not have begun this work without the early encouragement and amazing input from my wife, Susanne Cathryn Rohner, who has encouraged, supported, motivated, and sustained me throughout this effort, and who made it possible for me to make a home with her in a cedar cabin in the majestic silence of Lassen forest, where this book has been developed. And finally, infinite gratitude and love to all of my family, teachers, friends, and ancestors in this journey that we share.

Printed in the United States of America

David Bohm's Implicate Order:
A New Frontier in the Evolution of Consciousness

ISBN-13: 978-1-950761-04-3

TABLE OF CONTENTS

CHAPTER 1: INTRODUCTION 1

A NEW THEORY OF CONSCIOUSNESS 1

INTEGRAL METHODOLOGY: AN INTEGRAL PERSPECTIVE 3

Integral Philosophy and Integral Theory 4

CHAPTER 2: DAVID BOHM'S IMPLICATE ORDER 6

THEORY OF HIDDEN VARIABLES 19

Motion of "Particles" in Space–Time 21
Krishnamurti and Meditation 27

THE IMPLICATE ORDER 30

THE SUB-QUANTUM ORDER AND THE SUB-RELATIVISTIC LEVELS OF REALITY 34

HOLOMOVEMENT AND HOLOFLUX 46

Meaning, Form, and Information 50
Active Information and the Quantum Potential 53
The Sense of "I" and the Implicate Order 61

BOHM'S QUANTUM THEORIES AND CONSCIOUSNESS 63

Bohm's Waves Coming into and Going Out of a Point 65
Metaphor of the Quantum Black Hole 66
Metaphor of the Yin/Yang Symbol 67

CHAPTER 3: THE HOLOFLUX THEORY OF CONSCIOUSNESS 70

THE PLANCK CONSTANT ($\hbar$) 72

THE SUB-QUANTUM HOLOSPHERE 82

The Sub-Quantum Holoplenum 85
Wheeler's Self-Observing U 92
Creation and Dissolution of the Universe 95
Temperature/Energy of a Quantum Black Hole 98
Wheeler's "It from Bit" Black Hole 99

FIELDS, FREQUENCY, AND THE FREQUENCY DOMAIN 103

SINE WAVE AND FREQUENCY MOTION OF AN EXOPLANET 107

Information, Communication, and Meaning 119
Leonhard Euler and Imaginary Numbers 124
Derivation of Euler's Law 128
The Integral Operator and the Fourier Transform 133
The Real–Imaginary Domain and the Mandelbrot Set 138
Resonance in the Frequency Domain: Faraday Waves 143

Superposition in the Frequency Domain....149
Reality of Schrödinger Wavefunction (ψ)....152
HOLOFLUX IN THE HOLOPLENUM....154
The Dynamics of Holoflux....158
Holoplenum Plasma Display....159
Quantum Biology and the Holoflux Field....160
Quantum Brain Dynamics....161
HOLOFLUX ANALYSIS OF HUMAN RED BLOOD CELL....168
Dimensional Analysis and EMF Wavelength....171
Biosphere, Magnetosphere, Chromosphere....174
CHAPTER 4: SUMMARY AND CONCLUSIONS....185
THE HOLOFLUX THEORY OF CONSCIOUSNESS....186
STRATEGIES FOR TESTING THE HOLOFLUX HYPOTHESIS....192
UPLOADING CONSCIOUSNESS FROM WETWARE TO HARDWARE.201
REFERENCES....203
APPENDIX: THE HOLOFIELD THEORY OF ERVIN LÁSZLÓ....234

LIST OF FIGURES

Figure 1. Sub-quantum ocean floor metaphor. 40
Figure 2. Spectral domain and space–time domain. 42
Figure 3. The implicate/explicate order transform.............45
Figure 4. Active Information and the Quantum Wave. 57
Figure 5. Dualism vs. monism.. 62
Figure 6. Implicate/explicate as yin/yang. 67
Figure 7. Niels Bohr and his coat of arms. 68
Figure 8. David Bohm and Karl Pribram in Prague, 1992...71
Figure 9. Portrait of Max Planck, 1878... 74
Figure 10. A sub-quantum holosphere.. 84
Figure 11. Holoplenum formed by implicate order. 90
Figure 12. A regular three-dimensional tetrahedron...........91
Figure 13. Wheeler's self-observing U.. 94
Figure 14. Clock-cycle: digital holoflux. 96
Figure 15. Symbolic representation of the Bekenstein number of bits.. 101
Figure 16. James Clerk Maxwell at Trinity College. 104
Figure 17. A sine wave 107
Figure 18. Wavelength. 108
Figure 19. Water response to infrared radiation... 112
Figure 20. Atmospheric electromagnetic opacity. 113
Figure 21. Microtubules and capillaries as waveguides. 119
Figure 22. Mandelbrot: Origin of real–imaginary axis. 140
Figure 23. Faraday wave patterns......................................146
Figure 24. Resonance effect. 147
Figure 25. Sawtooth wave pattern created from five frequencies...151
Figure 26. Red blood cell...167
Figure 27. Isospheric capacity of red blood cell. 169
Figure 28. Supercomputer model of magnetosphere.. 175
Figure 29. Drawing of ventricular cavities.........................182
Figure 30. Microwave radar circuitry.. 184
Figure 31. Human electromagnetic wavelength. 194
Figure 32. Eight feasible bands for consciousness.. 199

By the same author:

The Little Book of Consciousness: Holonomic Brain Theory and the Implicate Order

Dr. Joye has not only created a map of consciousness; she has created a map of reality. Everything we call reality is experienced in consciousness, known in consciousness, and is in fact a modified form of consciousness. Reading this book will move you beyond physical materialism into better understanding the realities of consciousness.

——**DEEPAK CHOPRA, M.D.**,
Endocrinologist, founder of The Chopra Center, and author of *You Are the Universe* with the physicist Menas Kafatos

Tuning the Mind: The Geometry of Consciousness

Shelli Joye's integrative masterpiece, *Tuning the Mind*, reads like insane science fiction—quantum black holes, a universal center that's everywhere and nowhere, the dark energy holoflux, and of course, the mysterious Teilhardian isospheres. But the strange thing is, it's not insane, nor is it fiction. It's a novel and refreshing scientific take on the role of consciousness in the physical world. Well worth the read.

——**DEAN RADIN, Ph.D.**,
Chief Scientist, Institute of Noetic Sciences (IONS), and author of *Supernormal: Science, Yoga, and the Evidence for Extraordinary Psychic Abilities*

Chapter 1: Introduction

In this book the unique paradigm of the American physicist David Bohm is presented, a holoflux theory of an implicate order mapping the underlying energy dynamics of consciousness in the cosmos. Through application of an integral methodology, and maintaining congruence with widely held principles of physics, neurophysiology, and electrical engineering, a holoflux theory of consciousness is presented.[1] This paradigm is developed as an extension of the holonomic mind/brain research of Karl Pribram, the quantum potential theory of David Bohm, mapping the dynamics of consciousness as an evolving process of energy.

A New Theory of Consciousness

A new theory of consciousness is presented here as an extension of the theories pioneered by David Bohm and his collegue of many years, the Stanford brain scientist Karl Pribram. Core elements from their lifetime publications are examined and applied to construct a coherent, logical, and adequate model of consciousness that is in alignment with established principles of physics, mathematics, electrical engineering, and speculative philosophy.[2] The

---

[1] The integral methodology used in this book is described later in this Introduction.

[2] Whitehead, *Process and Reality*, 3; "Speculative philosophy is the endeavor to frame a coherent, logical, necessary system of general ideas in terms of which every

objective is to articulate a theory of consciousness that can be characterized as "the simplest possible theory that explains the data," a challenge put forth by the Australian philosopher of consciousness, David J. Chalmers, in an essay describing an "Outline of a Theory of Consciousness":

> It is not too soon to begin work on a theory. We are already in a position to understand certain key facts about the relationship between physical processes and experience and about the regularities that connect them . . . . Finally, the fact that we are searching for a *fundamental* theory means that we can appeal to non-empirical constraints such as simplicity and homogeneity in developing a theory. We must seek to systematize the information we have, to extend it as far as possible by careful analysis, and then to make the inference to the simplest possible theory that explains the data while remaining a plausible candidate to be part of the fundamental furniture of the world.[3]

Integral Methodology: An Integral Perspective

The holoflux theory, based upon Bohm's implicate order research, is here developed through application of an integral methodology that has been encouraged and developed within the California Institute of Integral Studies (CIIS) over the past forty

element of our experience can be interpreted."

[3] Chalmers, "Facing Up to the Problem of Consciousness," 217.

years. The Australian psychologist Jennifer Gidley, President of the World Futures Studies Federation, comments on the approach fostered at CIIS in her article on the need for an imminent evolution of consciousness:

> Although there is an emergence of new integrative material from within both science and the humanities, there is still a predominance of researchers who do not refer to scholarly research from neighboring disciplines . . . . An exception is the California Institute of Integral Studies. By contrast, integral thought is less explicit and tangible in Europe—with the term *transdisciplinary* apparently favored over *integral*.[4]

The methodology applied here is more than simply transdisciplinary. While it does include the effort to discern patterns and elicit connections from a set of normally disparate academic and scientific disciplines, the integral method fully acknowledges the importance of participatory, experiential dimensions, particularly in any approach to consciousness studies. Thus the theories of David Bohm are presented sequentially, in the context of his own unique life experiences. This methodology can be seen as a multiperspectival approach to the primary object of study, consciousness.

---

[4] Gidley, "Evolution of Consciousness as a Planetary Imperative," 201.

Integral Philosophy and Integral Theory

With ever-increasing information to assimilate, academic specializations have had to grow more focused. However, excessive focus and specialization has resulted in a human knowledge that is becoming increasingly fragmented. The archetypal pattern of the Tower of Babel comes to mind; each domain of knowledge has developed its own highly specialized language and conceptual framework, such that cross-departmental dialog has been impaired. According to the systems thinker, Edgar Morin, the problem seems to be fragmentation and lack of integration:

> An influx of knowledge at the end of the 20th century sheds new light on the situation of human beings in the universe. Parallel progress in cosmology, earth sciences, ecology, biology and prehistory in the 1960s and 1970s have modified our ideas about the universe, the earth, life and humanity itself. But these contributions remain disjointed. That which is human is cut up into pieces of a puzzle that cannot form an image . . . . The new knowledge, for lack of being connected, is neither assimilated nor integrated. There is progress in knowledge of the parts and paradoxical ignorance of the whole.[5]

The integral methodological approach in this book assumes that valid data may be found significantly beyond the traditional domains of scientific methodology. An integral method examines

[5] Morin, *Seven Complex Lessons in Education for the Future*, 21.

information from multiple and often disparate domains, seeking to perceive correlations between them, resonances among disparate domains of knowledge which might lead to conceptual bridges between these otherwise relatively isolated, compartmentalized knowledge.

Chapter 2: David Bohm's Implicate Order

Strong support for revolutionary new theories relating brain and consciousness can be found in the quantum physics of the American physicist David Joseph Bohm (1917–1992), a Professor Emeritus of Theoretical Physics at Birkbeck College, University of London. Bohm's early research career began at the University of California in Berkeley, where he completed his dissertation in plasma physics in 1944 under Robert Oppenheimer at the Lawrence Radiation Laboratory in Berkeley.[6]

Much as Karl Pribram's search for a mind/brain process that was missing in mainstream research approaches, Bohm's research was a focus on something that was missing in quantum theory, the relationship between matter and consciousness. Bohm derided physicists who seemed to show no interest in the *meaning* of the quantum world, those who used quantum mechanics merely as a mathematical tool for effectively predicting results in nuclear physics experiments. At the same time, Bohm perceives that reality lies beyond appearances, that it is the creative process of a whole, beyond theorizing, even given the repeatable appearances of scientific measurement:

> The quantum jump or quantum leap is a creative process . . . . I called it the implicate order because its the order of enfoldment that counts not the movement in a line. It manifests in the explicate. What we thought was the essence is now the

[6] Peat, *Infinite Potential.*

> appearance . . . . Namely, that the particle which was thought to explain the reality is now seen as an appearance. The essence of the true being is unknown. Even the implicate order is merely a concept, so even that should turn out to be an appearance. But we say be bringing in deeper more penetrating appearances we understand better, but we are never going to grasp the whole. Even a unified theory will only be an appearance.[7]

In his approach, Bohm stresses that the distinction between information and meaning is an important one, as did Karl Pribram (particularly in his *Languages of the Brain*, 1972). Unlike many physicists, whom he criticized for being information intoxicated, Bohm felt it of the utmost importance to search for meaning behind the information, and not to regard quantum mechanics simply as some exotic branch of calculation methodology. Understanding the underlying ontology behind quantum mechanics, Bohm felt, was a key to understanding the relationship between matter and consciousness, and here Bohm criticizes Bohr, the founder of quantum mechanics, who had won the Nobel Prize in 1913 for his quantum theory in modeling the hydrogen atom:

> Bohr's approach is to say nothing can be said about quantum mechanics at all other than to use it is a calculator, but my approach is to give it another appearance, more meaning. In the mechanical order it is hard to get much meaning.[8]

---

[7] Bohm, *Beyond Limits*, 8:00.

In the middle of his career as a theoretical physicist, Bohm's search for the underlying meanings to quantum theory was enriched through an introduction to a contemplative approach to the mind and consciousness expressed in the ideas of Jidu Krishnamurti. In 1961, while a professor of physics at Bristol University, Bohm attended a conference in London to hear a series of Krishnamurti lectures on perception, a topic that, as a quantum physicist, greatly interested Bohm. During the prior year, Bohm had become interested in theories of consciousness and perception expressed in the writings of philosophers and mystics, and had begun reading books on Yoga, Buddhism, and the work of the mathematician P. D. Ouspensky, a follower of the mystic G. I. Gurdjieff.[9] Public knowledge of his new interest resulted in further loss of credibility among Bohm's critics in the United States, as recounted here by his colleague and biographer, the physicist F. David Peat:

> When his former colleagues in the United States learned of this change of interest, it caused them considerable distress. In the years that followed, some lamented that Bohm had gone "off the rails," that a great mind had been sidetracked, and the work of an exceptional physicist was being lost to science.[10]

---

[8] Bohm, *Beyond Limits*, 12:55.

[9] Peat, *Infinite Potential*, 193–94.

[10] Peat, *Infinite Potential*, 194–95

Bohm, however, was completely serious in his search for an understanding of quantum theory, even though the quest had led him into these areas of philosophy, psychology, and mysticism. Bohm's driving interest can be seen in the first paragraph of *The Undivided Universe: An Ontological Interpretation of Quantum Theory*, completed in 1992, only weeks before he suffered a fatal heart attack; here he expresses his passion to understand the mysteries that continue to lie behind quantum theory, rather than regarding the theory simply as something "which we know how to use":

> The formalism of the quantum theory leads to results that agree with experiment with great accuracy and covers an extremely wide range of phenomena. As yet there are no experimental indications of any domain in which it might break down. Nevertheless, there still remain a number of basic questions concerning its fundamental significance which are obscure and confused. Thus for example one of the leading physicists of our time, M. Gell-Mann, has said "Quantum mechanics, that mysterious, confusing discipline, which none of us really understands but which we know how to use."[11]

To understand Bohm's approach to quantum theory it is useful to examine their development in the context of the complexity of his life. Born in Pennsylvania in 1917, Bohm took an early interest in science, having become an avid reader of science fiction by the age of seven. After graduating in physics and receiving the school's Mathematics Prize from the

[11] Bohm and Hiley, *The Undivided Universe*, 1.

University of Pennsylvania, he accepted a fellowship to enter graduate studies at Cal Tech. Within a year he found himself dissatisfied with the seemingly exclusive focus on problem solving at Cal Tech, a focus which seemed to discount Bohm's passion for seeking an understanding of what must be underlying ontological realities, favoring instead the development of mathematical models and their manipulation.[12] In 1941 Bohm met with J. Robert Oppenheimer at the University of California, Berkeley, having heard that in addition to being a brilliant theoretical physicist, the thirty-seven year old Oppenheimer encouraged wide ranging intellectual discussion among his grad students and postdoctoral researchers. Bohm soon after accepted a fellowship offer from Oppenheimer to study at Berkeley, and in 1941 Bohm moved from Los Angeles to begin graduate work in plasma research at Berkeley.[13]

After a year of intense collaboration with other physicists at the laboratory, often working more than sixty hours a week, Bohm gave his first presentation of the ideas he had thus far developed to model the complex nature of plasma, but early in the presentation, as he began to try to convey his new ideas to Oppenheimer and the rest of his committee, something unexpected occurred:

> A naturally shy person, as he began to speak he nonetheless felt that everything was going extremely well. Soon he had the sensation that he was going beyond physics

[12] Ibid., 31.

[13] Peat, *Infinite Potential*, 46.

> into something almost mystical, to the point where he felt himself in direct contact with everyone in the room. He was convinced that each individual consciousness had been transcended so that his audience was also sharing this experience. [14]

This intense feeling was overwhelming, leading immediately to a deep depression, and Bohm took a leave of absence, during which he was unable to work or study for almost an entire year.[15]

Even more problematical for Bohm during the first year of work on his dissertation under Oppenheimer was his growing interest in several activist antiwar groups, including the Committee for Peace Mobilization, the Campus Committee to Fight Conscription, and the Young Communist League, and "it was noted that Bohm had been involved in organizing activities for the Communist party and had distributed copies of Earl Browder's *Victory—and After*."[16] These associations and activities were later to haunt him.[17]

The Communist party in the United States had peaked at about 30,000 members in 1941, having grown steadily during the Great Depression due in part to the rapid growth of labor unions, but also because Communism was seen as a clear enemy of Fascism.[18]

---

[14] Ibid.

[15] Ibid., 47.

[16] Ibid., 62.

[17] Hiley and Peat, *Quantum Implications*.

In November 1942, Bohm joined the Communist party in Berkeley, but membership was not particularly exciting or eventful, as Bohm later described to the historian Martin Sherwin during an interview: "The meetings were interminable, discussing all these boring attempts to stir up things on the campus which didn't amount to much."[19]

Back at Berkeley in 1943, Bohm completed his doctoral research project, an effort to develop a mathematical theory with which to model the dynamics of plasmas, a poorly understood phenomenon that greatly intrigued Bohm:

> As he studied the plasmas he became struck by their extraordinary nature. They began to take on, for him, the qualities of living beings. When physicists studied a plasma by introducing an electrical probe, it would generate a charged sheath around the probe and neutralize its effects. It was as if the plasma were protecting itself and preserving its internal status.[20]

Bohm completed his project, and the results were considered by Oppenheimer to be spectacular; the calculations that Bohm developed in his thesis were able to model accurately the behavior of uranium fluoride plasma in an electric arc, and led directly to an elegant technique for separating U^{235} from U^{238}, a

---

[18] Peat, *Infinite Potential*, 57–58.

[19] Ibid., 58.

[20] Ibid., 66.

critical step in the Oppenheimer's effort toward building a nuclear bomb.[21]

Unfortunately, shortly before the publication of his dissertation, Bohm's previous political activities caught up with him. It was suddenly apparent that his affiliation with Communism made it impossible for him to obtain a security clearance, while paradoxically the importance of his own dissertation for the war effort led to it being considered highly classified:

> The scattering calculations (of collisions of protons and deuterons) that he had completed proved useful to the Manhattan Project and were immediately classified. But without security clearance, Bohm was denied access to his own work; not only would he be barred from defending his thesis, he was not even allowed to publish his own thesis.[22]

Oppenheimer was accordingly forced to petition the University of California Regents to rule on an exception to academic policy in order to approve Bohm's doctoral dissertation without allowing it to be published. The Regents assented and Bohm was awarded his doctorate and permitted to remain working at the Lawrence Laboratory, where he continued to focus his intense intellect upon the behavior of plasma flow and the mathematics of flux movement.[23] Here in an interview, Bohm describes his

---

[21] Ibid., 65.

[22] Ibid., 64.

[23] Plasmas constitute over ninety-nine percent of the matter

early interests and work with Oppenheimer during the period 1943–1946:

> When I worked at the Lawrence Laboratory, after taking my PhD, I became very interested in the electron plasma. This is a dense gas of electrons that exhibits radically different behavior from the other, normal states of matter and it was a key to much of the work the laboratory was doing at the time. My insights sprang from the perception that the plasma is a highly organized system, which behaves as a whole. Indeed, in some respects, it's almost like a living being. I was fascinated with the question of how such organized collective behavior could go along with the almost complete freedom of movement of individual electrons. [24]

Unfortunately, Bohm's early flirtation with Communism made the suppression of his dissertation only the first of what would eventually be a series of challenges to his academic career. Nevertheless, in 1947 the young physicist was hired as an assistant professor by Princeton University. During the relocation to New Jersey, he found a room near campus, and soon met his neighbor, another Princeton professor, the 57-year-old Albert Einstein, who, like Bohm, objected to quantum mechanics for its "lack of an objective description."[25] Bohm's clarity on this topic

---

of the universe—most stars and interstellar gases exist in this fourth state of matter (Peat, *Infinite Potential*, 65).

[24] Bohm and Peat, *Science, Order, and Creativity*, xiii.

was no doubt helped by the many discussions he had with Einstein in his days at Princeton.[26]

During his time at Princeton, Bohm also continued his research on the mathematics of plasma dynamics, while giving classes and public lectures on quantum mechanics, but he refused to accept the majority view which was, "and still is," in the words of his colleague, B. J. Hiley,

> that the precise nature of the conceptual changes is not important . . . . All that is needed is to work with the self-consistent mathematical formalism, which, in some mysterious way, correctly predicts the numerical results of actual experiments. After lecturing on the subject for three years, Bohm thought that this was not a satisfactory position to adopt so he decided to try to get a better understanding of the subject by writing a definitive textbook in which the physical aspects of the mathematics would be emphasized.[27]

Bohm worked furiously in his spare time on this major textbook that was to be a full interpretation of quantum theory, grounded both in mathematics and conceptual clarifications, which he hoped would resolve major unanswered problems within the quantum theory.[28] He also began working on a

---

[25] Peat, *Infinite Potential*, 88.

[26] Bohm, *The Special Theory of Relativity*, vi.

[27] Hiley and Peat, "The Development of David Bohm's Ideas," 4.

reinterpretation of quantum theory in a paper of his own, which he called the “hidden variable theory.” His textbook was published in 1951, and the paper in 1952, but in October 1951 Bohm fled the United States for Brazil, his reputation having been seriously affected by the politics of McCarthyism, and under the fear of being jailed for Contempt of Congress.[29]

In late 1949, while working on the final chapters of his book, Bohm had been served with a summons, a subpoena ordering him to appear before Senator Joseph McCarthy at the House Un-American Activities Committee. At the hearing, Bohm was asked to identify which of his previous numerous colleagues at Berkeley had been interested in Communism. Not wishing to incriminate his friends and colleagues, Bohm repeatedly had his attorney plead his “right to silence” under the Fifth Amendment to the United States Constitution. Notwithstanding, on December 4, 1950, Bohm was arrested at his home for Contempt of Congress, and though he made bail and was released, the administration at Princeton University peremptorily suspended him from “all teaching and other duties” for the duration of the trial.[30]

The publicity associated with the McCarthy hearings soon led to Bohm’s dismissal from his position at Princeton, and subsequent inability to find employment in the United States. With letters of reference from Einstein and Oppenheimer, Bohm was

---

[28] Bohm, *Quantum Theory*.

[29] Peat, *Infinite Potential*, 120.

[30] Peat, *Infinite Potential*, 98–99.

offered a position in the physics department of the University of São Paulo, the largest university in Brazil, then in the throes of development and expansion, during which had been seeking to attract a prominent quantum physicist from a mainstream American university.[31] Bohm decided to accept the position and prepared to make a hasty departure for Brazil, a day described here by his biographer:

> On the October day in 1951 when Bohm left the United States, thunderstorms and hurricane-force winds swept across New Jersey and New York. The flooded streets made it difficult for him to reach the airport . . . . On board the aircraft while taxiing to takeoff, he heard an announcement that the plane would return to the terminal because of an irregularity in the passport of one of the passengers. Already highly nervous, Bohm wondered if he was going to be arrested. The problem, however, concerned another passenger, who was removed from the flight.[32]

In December 1951, two months after his arrival in Brazil, the US consulate in São Paulo confiscated Bohm's passport, effectively stripping him of his US citizenship, and making it impossible for him to leave Brazil other than to return to the United States.[33]

---

[31] Baggott, *The Quantum Story*, 305.

[32] Peat, *Infinite Potential*, 120.

[33] Ibid., 124.

Undeterred, with his passion for learning undiminished, Bohm continued to work, and he was soon fluent in self-taught Portuguese, sufficiently fluent to lecture at the University of São Paulo.[34] He was especially encouraged by favorable peer reviews of his recent US publication of *Quantum Theory*, a 628-page textbook interpreting quantum theory, and which begins in the first paragraph to emphasize the importance of familiarity with Fourier analysis:

> Modern quantum theory is unusual in two respects. First, it embodies a set of physical ideas that differ completely with much of our everyday experience, and also with most experiments in physics on a macroscopic scale. Second, the mathematical apparatus needed to apply this theory to even the simplest example is much less familiar . . . . It seems impossible, however, to develop quantum concepts extensively without *Fourier analysis*. It is, therefore, presupposed that the reader is moderately familiar with Fourier analysis.[35]

Theory of Hidden Variables

But perhaps the greatest significance of *Quantum Theory* lies in the fact that Bohm presents, once again, but this time in a major text book, his theory of hidden variables: "Perhaps there are hidden variables that really control the exact time and place of a transfer of a quantum, and we simply haven't found

[34] Ibid., 126.

[35] Bohm, *Quantum Theory*, 1. Italics added.

them yet."[36] Einstein's major objection to the quantum theory was its almost complete lack of an objective description, and it was this that Bohm addressed in his new interpretation.[37]

At Princeton, Bohm and Einstein had agreed that there must be underlying processes yet to be discovered. Quantum mechanical calculations, while they work in the prediction of experimental evidence, treat particle actions collectively; in a similar way as taking temperature measurements are used to predict the motion of billions of gas molecules. In neither case can actual particle motion be mapped or predicted at the real dimensions being examined. Einstein and Bohm felt there must be principles that affect motion at a lower scale than that addressed by the collective mathematics of quantum mechanics (or temperature measurements).

> This assumption has been the object of severe criticisms, notably on the part of Einstein, who has always believed that, even at the quantum level, there must exist precisely definable elements or dynamical variables determining (as in classical physics) the actual behavior of each individual system, and not merely its probable behavior . . . . The suggested interpretation provides a broader conceptual frame-work than the usual interpretation, because it makes possible a precise and continuous description of all process, even at the quantum level. This

[36] Ibid., 29.

[37] Peat, *Infinite Potential*, 86.

> broader conceptual framework allows more general mathematical formulations of the theory than those allowed by the usual interpretation. Now, the usual mathematical formulation seems to lead to insoluble difficulties when it is extrapolated into the domain of distances of the order of 10^{-33} cm or less. It is therefore entirely possible that the interpretation suggested here may be needed for the resolution of these difficulties.[38]

In his paper on hidden variables can be seen Bohm's early fascination with the physics of the greatest depths of scale, far below the normal range at which even quantum mechanics is applied, down at distances of the Planck length (approximately 10^{-33} cm), a scale at which "classical notions such as causality or distance between events cannot be expected to be applicable."[39] The Planck constant is discussed in detail in Chapter 5.

Motion of "Particles" in Space–Time

Bohm explains apparent linear motion with an ontology derived from quantum theory when he speculates that, rather than manifesting as a solid, continuous particle in space–time, "the proton takes the form of a wave that collapses inward from all space," like a whirlpool in water.[40]

---

[38] Bohm, "Interpretation of the Quantum Theory in Terms of 'Hidden Variables,'" 166.

[39] Garay, "Quantum Gravity and Minimum Length," 145.

Similarly, in a conclusion that is contrary to the conventional assumption that electrons and protons are well-defined particles, Bohm describes the apparent motion of protons and electrons as:

> a continuous process of inward collapse and outward expansion. Therefore, every elementary particle collapses inward from the whole of space. In fact, each elementary particle is a manifestation of the whole universe.[41]

While his textbook on *Quantum Theory* was generally well received, and subsequently used in undergraduate courses on quantum mechanics worldwide for its clarity of language, analogy, and supporting calculus, the concept of *hidden variables*, described in the book and in his 1952 paper published exclusively on the "hidden variables" was pounced upon almost universally by quantum physicists, as being incompatible with the prevailing "Copenhagen interpretation."[42] In Copenhagen, during 1925 and 1927, Niels Bohr and two of his young students, Heisenberg and Pauli, at a series of meetings and conferences, both formal and informal, had developed what was later to be called the Copenhagen Interpretation of quantum mechanics, a conjectural descriptive explanation of what the working

---

[40] Peat, *Infinite Potential*, 48.

[41] Peat, *Infinite Potential*, 48–49.

[42] Bohm, "Interpretation of the Quantum Theory in Terms of 'Hidden Variables.'"

mathematical equations seemed to imply.[43] To most physicists, having worked under general agreement for over 20 years under the Copenhagen interpretation, Bohm's hidden variable theory was seen as being almost heretical, a direct challenge to the Copenhagen interpretation; yet from Bohm's point of view, it was the Copenhagen approach, with its heavy reliance on mathematical probability theory, that was random.[44] In this, Albert Einstein, Bohm's former neighbor, friend, and mentor at Princeton, can be seen in full agreement; Einstein is quoted here in a 1926 letter to Max Born, where he comments on the recently published Copenhagen interpretation, and expresses special concern regarding the deep reliance of probability theory in quantum mechanics:

> Quantum mechanics is certainly imposing. But an inner voice tells me that it is not yet the real thing. The theory says a lot, but does not really bring us any closer to the secret of the "old one." I, at any rate, am convinced that He does not throw dice.[45]

The prevailing consensus, in accord with the Copenhagen interpretation, holds that quantum events are indeterminate and discontinuous, and therefore the physics of process is not linearly predictable at quantum scales of matter, and that only probability mathematics works in predicting outcomes. Bohm

---

[43] Peat, *Infinite Potential*, 51.

[44] Ibid., 52.

[45] Born, *The Born–Einstein Letters*, 91.

opposed this axiom with his conviction that the process of the universe cannot be attributed to purely random statistical activity, but that there is an underlying order and comprehensive connectivity, a unity to the cosmos, a wholeness not predicted by mere probability theory.[46]

In his hidden variable theory, however, Bohm assumes the universe to be coherent and predictable, and considers the apparent randomness to be misleading, proposing instead that there must exist hidden variables operating at sub-quantum levels, hidden forces and processes operating at lower levels, below the currently assumed lower limit of matter, and not yet accessible with current instrumentation. Understandably, Bohm's new theory of hidden variables was immediately rejected by the community of physicists, in part for political reasons (Bohm's earlier communist affiliations), but primarily due to the firm attachment to the consensual acceptance of the Copenhagen interpretation and the predictable results proven by use of quantum mechanical calculations.[47]

At best, Bohm's new theory was ignored, but some went so far as to accuse him of being a Trotskyite and a traitor; one physicist termed Bohm's work "juvenile deviation," and another called Bohm "a public nuisance."[48] This strong reaction against Bohm's *hidden variable* hypothesis even led his own one-time supervisor, mentor, and hero, Robert Oppenheimer, to

---

[46] Peat, *Infinite Potential*, 108.

[47] Ibid., 148.

[48] Ibid., 133.

gather a meeting of physicists at Berkeley in 1953 with the objective of disproving Bohm's theory; the meeting closed with Oppenheimer being heard to say, "If we cannot disprove Bohm, we must continue to ignore him."[49]

In 1955, Bohm left Brazil to accept a position as a physics professor at the Israel Institute of Technology in Haifa, where he meets his future wife, Saral, an artist. In 1957 Bohm and Saral left Israel where Bohm accepted a position at Bristol University in England. In London, that same year, Bohm presented his mature version of the hidden variable concepts, which he had first begun to develop in 1952 in *Quantum Theory*. In his London paper Bohm introduced the idea of processes occurring at sub-quantum levels, operating at considerably smaller dimensions than those at which quantum mechanical calculations modeled the activity of atomic particles. This was the beginning of Bohm's conceptualization of the implicate order, through considering the possible ontological reality of dimensions far below the quantum dimensions being explored by quantum mechanics.

During his second year in England, Bohm presented a series of lectures at London University detailing his hidden variable theory; the lecture was attended by a young doctoral student in physics, John Bell, who had become interested in mathematical approaches to the nonlocality problem in quantum mechanics. Bell, encouraged by Bohm's theories, went on to develop a mathematical proof of nonlocality, a phenomenon that has since been verified several times

---

[49] Ibid.

in rigorous experiment.[50] It is of interest to note that Bell said, of Bohm's theory, "No one can understand this theory until he is willing to see psi as a real objective field rather than just a probability amplitude."[51]

Bohm did not limit his intellectual pursuit exclusively to physics. During his three years at Bristol University he began to read widely in areas of speculative philosophy and soon became acquainted with the process philosophy of Alfred North Whitehead, the Cambridge mathematician turned Harvard philosopher. One evening, according to Bohm, while reading *Process and Reality*, he experienced what seemed to be an epiphany, a visualization of infinity in the image of a large number of silvered spherical mirrors, each one reflecting all of the others, a cosmos composed of an infinity of reflections, and of reflections of reflections: "Every atom was reflecting in this way, and the infinity of these reflections was reflected in each thing; each was an infinite reflection of the whole." [52] The influence of this image can be found 20 years later in Bohm's understanding of what he had by then come to call, "the Implicate Order." The image of mirrors reflecting the whole is echoed in *Wholeness and the Implicate Order* in the following passage, which describes Bohm's concept of the holomovement:

---

[50] Ibid., 168.

[51] J. T. Bell, *The Speakable and Unspeakable in Quantum Mechanics*, 128.

[52] Peat, *Infinite Potential*, 186.

> In certain ways this notion is similar to Leibniz's idea of monads, each of which "mirrors" the whole in its own way, some in great detail and others rather vaguely. The difference is that Leibniz's monads had a permanent existence, whereas our basic elements are only moments and are thus not permanent.[53]

At Bristol in the early 1960s, during his wide-ranging reading in philosophy and Eastern metaphysics, Bohm was introduced to the writings of the Russian mathematician P. D. Ouspensky, who had written several books on the metaphysics of the mystic G. I. Gurdjieff.[54] In 1962 Bohm managed to meet in London with J. G. Bennet, himself a mathematician as well as an authority on Gurdjieff; the two corresponded over the next several years, but "their connection was broken," as Bohm's biographer says, in a footnote:

> Bohm became increasingly critical of Gurdjieff's emphasis upon the essential separation of an observing consciousness. For Bohm there was a "total awareness" that lies beyond consciousness and is the field in which creativity operates.[55]

Krishnamurti and Meditation

It was also while at Bristol University that Bohm first became acquainted with the philosopher Jidu

[53] Bohm, *Wholeness and the Implicate Order*, 207.

[54] Ouspensky, *In Search of the Miraculous*.

[55] Peat, *Infinite Potential*, 213.

Krishnamurti. In the Bristol public library, Bohm's wife, Saral, discovered a book in the philosophy section discussing the topics of perception and thought, and brought it home as of possible interest. Bohm was immediately fascinated by what Krishnamurti had to say:

> My first acquaintance with Krishnamurti's work was in 1959 when I read his book *First and Last Freedom.* What particularly aroused my interest was deep insight into the question of the observer and the observed . . . I felt that it was urgent for me to talk with Krishnamurti directly and personally as soon as possible. And when I first met him on one of his visits to London, I was struck by the great ease of communication with him, which was made possible by the intense energy with which he listened and by the freedom from self-protective reservations and barriers with which he responded to what I had to say . . . it was in essence of the same quality as that which I had met in contacts with other scientists with whom there had been a very close meeting of minds. And here, I think especially of Einstein who showed a similar intensity and absence of barrier in a number of discussions that took place between him and me. After this, I began to meet Krishnamurti regularly and to discuss with him whenever he came to London.[56]

In 1961, Bohm left Bristol University to accept the offer of Chair of Theoretical Physics, offered to him

[56] Bohm and Krishnamurti, *The Limits of Thought,* vii.

by J. D. Bernal, the head of the physics department (and a dedicated Marxist) at Birkbeck College, University of London. In that same year Bohm, for the first time, arranged a private conversation with Krishnamurti in London. The two continued their dialogues, at first privately, and then eventually in public venues, before audiences in London, Switzerland, and later in Ojai, California.[57]

In Switzerland in 1965, the first public recordings were made of a series of dialogues between Krishnamurti and Bohm. Undoubtedly Bohm's previous reading experience of Whitehead, Hegel, Ouspensky, and Indian philosophy had prepared him well for these flowing dialogues with Krishnamurti, and likely influenced his own conceptualizations. In addition to published books containing transcripts of many of these dialogues, numerous video and audio recordings have been released; a recent YouTube search on "Krishnamurti" and "Bohm" pulls up 32,000 results.[58]

It was in 1967, during the "Summer of Love," that Bohm and his wife, at the urging of Krishnamurti, both became vegetarians and took up the daily practice of meditation. Bohm was particularly interested in Krishnamurti's teaching on the process of "dying to thought" in which the physical brain no longer gives energy to the movement of thought process in the brain.[59] According to Krishnamurti, it is this incessant

---

57 Peat, *Infinite Potential*, 213.

58 Bohm and Krishnamurti, *The Ending of Time*; and Bohm and Krishnamurti, *The Limits of Thought*.

conditioned mental thought process that masks and conceals the silence within which the less conditioned consciousness process might be perceived, and within which new modes of "non-verbal thought" might operate. Bohm here describes his own understanding of Krishnamurti's teaching on the process of meditation:

> Krishnamurti has observed that the very act of meditation will, in itself, bring order to the activity of thought without the intervention of will, choice, decision, or any other action of the "thinker." As such order comes, the noise and chaos which are the usual background of our consciousness die out, and the mind becomes generally silent . . . . In this silence, Krishnamurti says that something new and creative happens, something that cannot be conveyed in words, but that is of extraordinary significance for the whole of life. So he does not attempt to communicate this verbally, but rather, he asks of those who are interested that they explore the question of meditation directly for themselves.[60]

It is easy to imagine how Bohm's exceptional, trained capability (as a theoretical physicist) for focus on abstractions served to reinforce his growing capabilities in the practice of formless meditation. Bohm soon established a daily period in the early morning during which he would walk slowly in the park or countryside while watching the movement of

[59] Peat, *Infinite Potential*, 227–28.

[60] Bohm and Krishnamurti, *The Limits of Thought*, vii.

this thoughts, not allowing himself to become distracted or "caught up" in their content; this practice continued throughout his life.[61]

The Implicate Order

It is of interest to note that within two years of beginning his daily meditation walk, Bohm conceptualized a significantly new idea, the "implicate order," which ontologically strengthened his hidden variables theory. By 1969 the subjects of topology, nonlocality, and pre-space had come to the forefront of discussion within the physics department at the University of London, and as his colleague, the physicist B. J. Hiley recalls, Bohm had a breakthrough in his understanding: "Suddenly, a new idea emerged from Bohm, apparently entirely out of the blue . . . Bohm called it the *implicate order*."[62]

Bohm identified this implicate order as being the source of the hidden, sub-quantum variables which he hypothesized were providing the causality underlying the operation of quantum mechanics, and he began an effort to model the effect of the implicate order mathematically.[63] In 1925, Werner Heisenberg had taken a similar view:

> Heisenberg, who at age twenty-four, was young even by physics standards, tried to save classical mechanics by abandoning it at

[61] Peat, *Infinite Potential*, 229.

[62] Peat, *Infinite Potential*, 256.

[63] Ibid.

> Nature's bottom rung. Inside the atom, he declared, not only do particles and electron orbits have no meaning, but neither do even such basic classical properties as position, momentum, velocity, and space and time. And because our imaginations require a space–time container, this atomic world cannot be easily pictured.[64]

But Heisenberg's approach fell by the wayside when the Copenhagen approach proved so successful in predictive calculations of nuclear particle trajectories using probabilities. Decades later, Bohm pursued alone a similar idea, and in 1980 published *Wholeness and the Implicate Order*, in which is revealed both the mathematics and a mature conceptual model for his proposed implicate order, a model based not on "things" considered as objects, nor restricted solely to phenomena in a Cartesian space–time order, but rather modeling a lens-like flowing of simultaneous enfolding and unfolding dimensions between two orders, an implicate order and an explicate order, as Bohm explained:

> There is the germ of a new notion of order here. This order is not to be understood solely in terms of a regular arrangement of *objects* . . . . Rather, a *total order* is contained, in some *implicit* sense, in each region of space and time. Now, the word "implicit" is based on the verb "to implicate." This means, "to fold inward." So we may be led to explore the notion that in

---

[64] Crease, *The Great Equations*, 220.

> some sense each region contains a total structure "enfolded" within it.[65]

Bohm went so far as to describe the implicate order as the ground within which the entire universe is enfolded at each and every "point" in space–time, as the Finnish philosopher Paavo Pylkkänen, a former student and friend of Bohm's, explains clearly:

> Just think of all the atoms and particles that constitute your body. We are used to thinking about them as tiny little things that just passively sit there. But quantum field theory, as interpreted by Bohm, suggests otherwise. There is a sense in which each particle in your body enfolds information about the whole universe . . . . There is also a sense in which information about each particle in your body is enfolded throughout the universe.[66]

However, Bohm's focus of interest clearly ranged beyond the conventional boundaries of physics, into theories of mind, thought, and consciousness. It is easy to conclude that Bohm's dialog with Krishnamurti influenced the expansion of this focus, bringing to the forefront for Bohm issues of thinking and mind. But with his deep understanding of quantum theory, Bohm was able to see a direct connection between thought, consciousness, and the implicate order, as can be discerned in his description of the thinking process, recorded in a 1990 interview:

---

65 Bohm, *Wholeness and the Implicate Order*, 149.

66 Pylkkänen, *Mind, Matter, and the Implicate Order*, 21.

> And that's how the thinking process goes, it's enfolded in your consciousness, it unfolds to a certain thought, folds back, and then the next thought appears, different; a series of thoughts not too different seems to be continuous.[67]

In imagining this process as an "enfoldment," we are led to the obvious next question, "Enfoldment into what?" And Bohm's non-Cartesian answer then, is enfoldment *into something within*, enfoldment into the implicate order. How then might this be conceptualized? Imagine the universe at the lowest possible spatial range. Moving to ever-smaller dimensions, to the very bottom of the dimensional scale, we encounter the end of space, according to accepted quantum theory, at a limiting length, below which length has no meaning.[68] Thus this implicate order, located at the center, everywhere, can be seen as the ground, base, or center out of which dynamically springs the space–time continuum itself, the explicate order.

The Sub-Quantum Order and the Sub-Relativistic Levels of Reality

Bohm applied the term "sub-quantum order" to that order of physics which he viewed as expressing the underlying causal reality *from which* quantum mechanics operates, calling this approach "going beyond the quantum theory."[69] While quantum

---

[67] Bohm, *Beyond Limits*, 8:00.

[68] As discussed shortly, this is the Planck length of 10^{-33} cm.

mechanical theory works well for predicting the path of nuclear particles in dimensional ranges around 10^{-15} meters, the theory places a heavy insistence on pure probability considerations, and this randomness had always bothered Bohm. As early as 1952, in his textbook, *Quantum Theory,* Bohm states:

> We have come to the point of view that the wavefunction is an abstraction, providing a mathematical reflection of certain aspects of reality, but not a one-to-one mapping.[70]

He further argues that the ontological cause is to be discovered at lower levels, below quantum mechanical dimensions, and that "quantum theory is inconsistent with the assumption of hidden causal variables" governing processes at the lowest, sub-quantum levels.[71] At the end of the Introduction to his 1993 book, Bohm predicts the failure of the "current laws of physics" at the Planck length:

> The idea is that there will be a stochastic sub-quantum and sub-relativistic level in which the current laws of physics will fail. This will probably first be encountered near the Planck length of 10^{-33} cm . . . . This idea is connected with our ontological interpretation by means of a model of a particle as a sequence of incoming and outgoing waves, with successive waves very close to each other . . . . One of the main new

[69] Bohm and Hiley, *The Undivided Universe*, 9.

[70] Bohm, *Quantum Theory*, 622

[71] Ibid.

> ideas implied by this approach is that the geometry and the dynamics have to be in the same framework, i.e. that of the implicate order.[72]

This leads us to the observation that there exist three sets of framework-dependent mathematical models, applicable in three ranges: (1) Newtonian mechanics, operating at human dimensions, in the 10^0 meter range, (2) quantum mechanics, operating at nuclear particle dimensions, in the 10^{-15} meter range, and (3) Bohm's "hidden variables," operating at the boundary of space–time, in the 10^{-35} meter range. Here Bohm reiterates an observation he had previously made regarding the important significance of the dimension 1.616 x 10^{-33} cm, the Planck length, and designated with the symbol l_P in quantum theoretical calculations.[73] Over a decade earlier, in a 1980 publication, *Wholeness and the Implicate Order*, Bohm, in discussing the calculation of what is called "the zero-point energy" for point in space, provides another explanation for the significance of this 10^{-33} cm distance:

---

[72] Bohm and Hiley, *The Undivided Universe*, 9–10.

[73] See also Chapter 5, "The Planck Constant"; the Planck length of 10^{-33} cm. was first discussed by the German physicist Max Planck, the 1918 Nobel Prize winning originator of quantum mechanics, and is derived from three values: (1) the speed of light in a vacuum, (2) the gravitational constant, and (3) the Planck constant, *h*, which is required to calculate quantum changes of wavelength as a function of temperature (Bruskiewich, *Max Planck and Black-Body Radiation*).

> We come to a certain length at which the measurement of space and time becomes totally indefinable. Beyond this, the whole notion of space and time as we know it would fade out, into something that is at present unspecifiable . . . . When this length is estimated it turns out to be about 10^{-33} cm. This is much shorter than anything thus far probed in physical experiments (which have gone down to about 10^{-17} cm or so). If one computes the amount of energy that would be in one cubic centimeter of space, with this shortest possible wavelength, it turns out to be very far beyond the total energy of all the matter in the known universe.[74]

It is here then, in the region of the Planck length at 10^{-33} cm, that Bohm predicts that there will be found *a boundary* separating an outer, *explicate order* from an inner, *implicate order*. There will also be seen gaps of, at minimum, 10^{-33} cm between each and every concentric shell that might be surrounding the central area, a spherical locus of Planck length diameter l_P. This is in accord with the founding quantum theory of Max Planck, first published in 1900, in which he proposed that a photon could have only distinct, fixed energy states, separated by integers, which he termed "quanta."[75] Bohm uses such quantum theoretical considerations to point out a fallacy still held by many, the classical Cartesian assumption that space is continuous:

[74] Bohm, *Wholeness and the Implicate Order*, 190–91.

[75] Bohm, *Quantum Theory*, 18.

> What of the order between two points in space? The Cartesian order holds that space is continuous. Between any two points, no matter how close they lie, occur an infinite of other points. Between any two neighboring points in this infinity lies another infinity and so on. This notion of continuity is not compatible with the order of quantum theory . . . . Thus the physicist John Wheeler has suggested that, at very short distances, continuous space begins to break up into a foam-like structure. Thus the "order between" two points moves from the order of continuity to an order of a discontinuous foam.[76]

According to Bohm, at the ultimate bottom level of these subdivided infinities, between two points in space, like an end-of-stop bumper terminating a railroad track, will be found a cosmologically fixed boundary at the Planck length of 10^{-33} cm, beyond which no further subdivision is meaningfully possible. It is here, at this smallest of possible spatial coordinates, that we encounter the origin of the dimension of space that finds its maximum in the diameter of the currently expanding universe (the bandwidth continually increases, at the speed of light).

Viewed another way, to use the depths of the ocean as an analog or metaphor, Bohm's sub-quantum mechanics can be viewed as being at the bottom of a dimensional ocean with a floor significantly below the depths of the region mapped and explored by the mathematics of quantum mechanics. The image in Fig. 9 plots three ranges of dimensional scale along a

[76] Bohm and Peat, *Science, Order, and Creativity*, 311–12.

vertical axis. On the same vertical axis can be seen dimensional ranges within which operate (1) classical Newtonian physics, (2) Quantum physics, and (3) Bohmian physics, each operating within distinct ranges, separated by over 10^{15} orders of magnitude. In a real sense, Bohm was plumbing this depth of dimensional physics, far below the level at which conventional quantum theory is applied, at depths of scale considerably below those of the dimensions of a Monarch butterfly wingspan, a uranium atom nucleus, a proton, or a calculated electron (photon) diameter. Bohm comments on the depths of this range:

> It is interesting to note that between the shortest distance now measurable in physics (10^{-16} cm) and the shortest distance in which current notions of space–time probably have meaning (10^{-33} cm), there is a vast range of scale in which an immense amount of yet undiscovered structure could be contained. Indeed this range is roughly equal to that which exists between our own size and that of the elementary particles.[77]

[77] Ibid., 86.

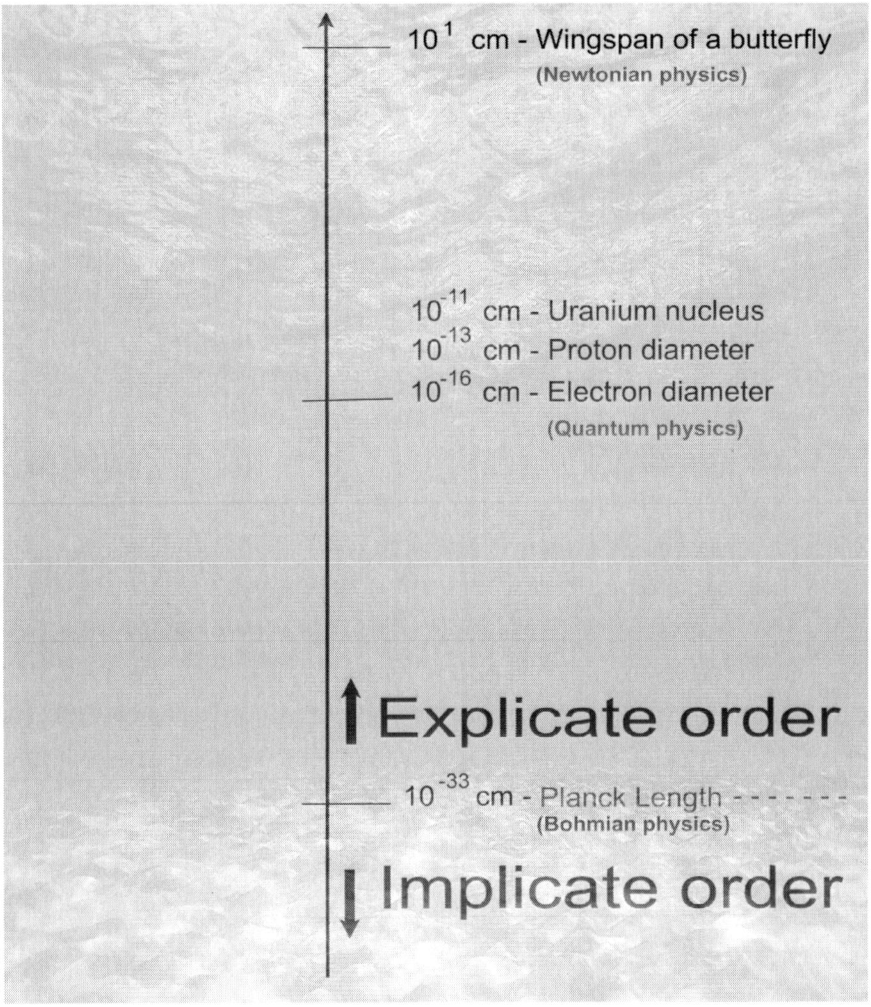

Figure *1*. Sub-quantum ocean floor metaphor.[78]
Graphic by author.

Note that quantum mechanics is applicable in a range only halfway down between the scale of the human biosphere and the depths of space–time at the Planck length. The mathematical models (and

[78] Data from Benenson, Harris, Stocker, and Holger, *Handbook of Physics*.

operational physics) that work in one range (for example, Newtonian mechanics as it operates within the human body parts) are in general not applicable within the other ranges (such as the operational physics of quantum mechanics working in electron and proton ranges).

At the absolute bottom of the figure's ocean floor lies the Planck length of 10^{-33} cm., above which the explicate order manifests in increasingly larger ranges of scale until the dimensional length of a butterfly (about 1 cm) is reached, floating on the surface of the explicate order's ocean depth of scale. Of course, a more inclusive scale would expand upwards to include the size of the universe, with a diameter of 10^{26} meters, and the human dimensional range would then be seen to lie approximately half way between these two limits, the size of the universe (10^{26} m) and the Planck length (10^{-35} m).[79] Such a complete scale range, spanning the entire cosmos, is depicted in Fig. 10.

---

[79] Gott et al., "A Map of the Universe."

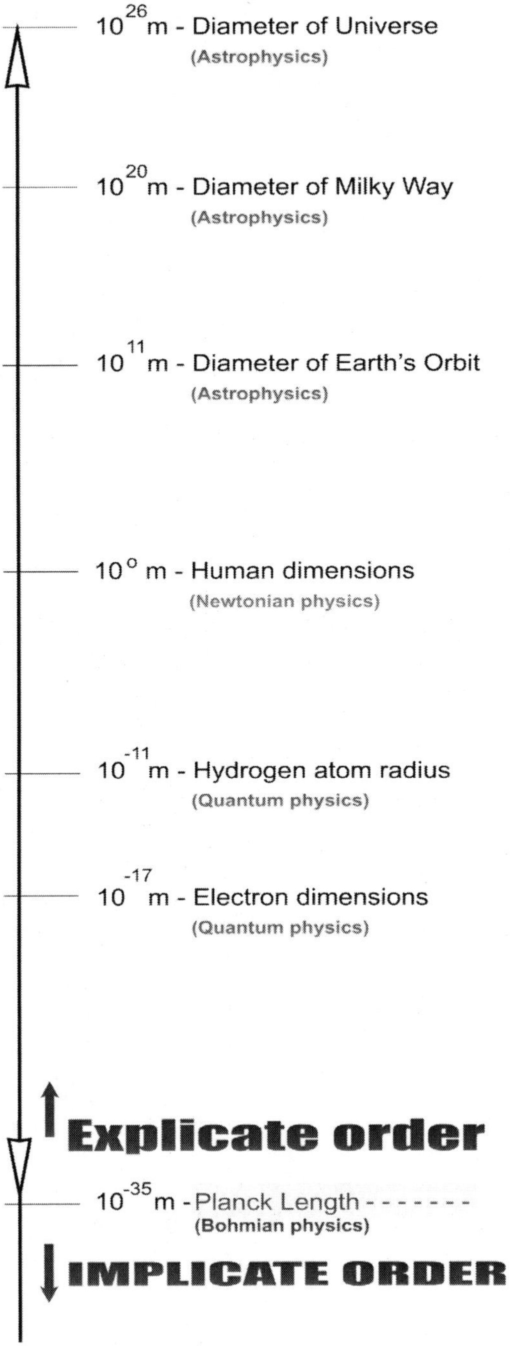

Figure 2. Spectral domain and space–time domain. Graphic by author.

Another foundation upon which Bohm's model of the universe grew can be seen in his vision of an endless infinity of ordered fields, as he describes here:

> We may suppose that the universe, which includes the whole of existence, contains not only all the fields that are now known, but also an indefinitely large set of further fields that are unknown and indeed may never be known in their totality. Recalling that the essential qualities of fields exist only in their movement we propose to call this ground the *holomovement.* It follows that ultimately everything in the explicate order of common experience arises from the holomovement. Whatever persists with a constant form is sustained as the unfoldment of a recurrent and stable pattern, which is constantly being renewed by enfoldment, and constantly being dissolved by unfoldment. When the renewal ceases the form vanishes.[80]

Fields need not exist only within space–time. Einstein's theory of general relativity posits ten fields to describe the movement of gravity in four dimensions.[81] As early as 1919 a German mathematician, Theodor Kaluza, sent Einstein a paper proposing an extremely small fifth dimension in which additional frequency and phase fields could provide a solution to resolve Einstein's struggle to resolve the connection between gravity and electromagnetism.[82] In

---

[80] Bohm and Hiley, *The Undivided Universe*, 357.

[81] Yau and Nadis, *The Shape of Inner Space*, 10.

1926 the Swedish physicist Oskar Klein "drawing on quantum mechanics, calculated the size of this compact dimension, arriving at a number that was tiny indeed—around 10^{-30} cm in circumference."[83] The theory became known as the Kaluza-Klein theory, offering the potential of extra fields in an additional dimension as a promising approach to resolving Einstein's search to connect gravity and electromagnetism. Einstein wrote to Kaluza, saying he liked the idea "enormously" and in fact continued working with the theory for the next 20 years; ultimately, however, the Kaluza-Klein theory was discarded because of its prediction of a particle so small that it could not be proven to exist using any foreseeable technology.[84]

But by the same token it has also been impossible to *disprove* a cosmological reality existing at such small dimensions, and it is here, at the limiting space–time scale of 10^{-33} cm, that Bohm hypothesizes an implicate order.[85] In Fig. 11, Pribram's diagrammatic concept of the Fourier transform is depicted with Bohm's implicate order to the left, in the spectral domain, and Bohm's explicate order to the right, in the space–time domain.

---

[82] Ibid., 10–11.

[83] Ibid., 12.

[84] Ibid., 13.

[85] Bohm, *Wholeness and the Implicate Order*, 190.

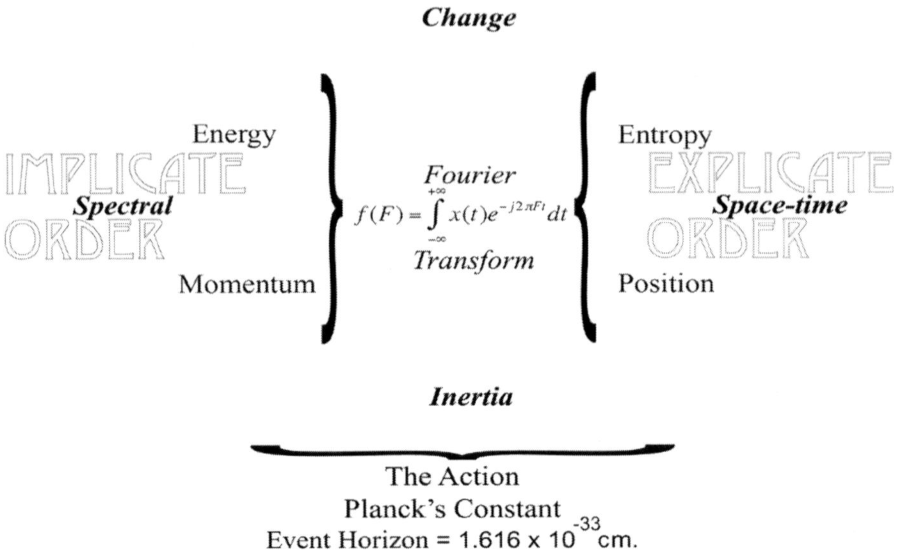

Figure 3. The implicate/explicate order transform. Graphic by author.

In the diagram, the process linking the two orders is shown as the Fourier transform function, implementing a reciprocal flow of spectral information between the implicate order into the explicate order, a two-way process, operating at an event horizon of the Planck limit of 10^{-33} cm, a process which Bohm terms the "holomovement," and which, he tells us, is grounded in the implicate order:

> The implicate order has its ground in the holomovement which is vast, rich, and in a state of unending flux of enfoldment and unfoldment, with laws most of which are only vaguely known, and which may even be ultimately unknowable in their totality . . . . Nevertheless, the overall law (holonomy) may be assumed to be such that in a certain sub-order, within the whole set of the implicate order, there is a totality of

> forms that have an approximate kind of recurrence, stability, and separability. Evidently these forms are capable of appearing as the relatively solid, tangible, and stable elements that make up our "manifest world."[86]

Holomovement and Holoflux

In 1978 Bohm discusses these concepts with remarkable clarity during a Harvard conference on "Science and Mysticism: Exploring the New Realities." After his presentation, which he called, "The Implicate Order: The Holodynamic Model of the Universe," Bohm was interviewed by Renée Weber for *ReVision* magazine, and here he expresses the difference between holography and the holomovement, and introduces the term "holoflux":

> **Weber**: Could we begin by clarifying the difference between the holomovement, the holograph and the implicate order?
>
> **Bohm**: Holomovement is a combination of a Greek and Latin word and a similar word would be holokinesis or, still better, *holoflux*, because "movement" implies motion from place to place, whereas flux does not. So the *holoflux* includes the ultimately flowing nature of what is, and of that which forms therein. The holograph, on the other hand, is merely a static recording of movement, like a photograph: an abstraction from the holomovement. We therefore cannot regard the holograph as

---

[86] Ibid., 186.

> anything very basic, since it is merely a way of displaying the holomovement, which latter is, however, the ground of everything, of all that is. The implicate order is the one in which the holomovement takes place, an order that both enfolds and unfolds. Things are unfolded in the implicate order, and that order cannot be entirely expressed in an explicate fashion. Therefore, in this approach, we are not able to go beyond the holomovement or the *holoflux* (the Greek word might be holorhesis, I suppose) although that does not imply that this is the end of the matter.[87]

In a theory congruent with Pribram's, Bohm hypothesizes that consciousness must operate dynamically in a holonomic fashion, driven by a quantum potential wave, and that the information guiding motion in space–time must originate from an *other order*, the implicate order, which lies outside of the commonly perceived explicate order of the space–time domain.[88] From early in his career, Bohm had tried to visualize the motion of quantum "particles," not moving in linear sequential tracks as viewed in a cloud chamber, but as oscillating in and out of space–time.[89]

Bohm, here in a 1990 interview recorded for Dutch television, not only describes this movement or

---

[87] Weber, "The Physicist and the Mystic," 187. Italics added.

[88] Hiley and Peat, eds. *Quantum Implications: Essays in Honour of David Bohm.*

[89] Ibid.

"holomovement" as superseding the so-called "particle" phenomena of quantum mechanics, but in doing so, Bohm emphasizes the reality of the whole, the continuum of mind and matter, and discusses the general implications of this for perception:

> Bohm: Now, I saw that you could understand the quantum mechanics in terms of that process. That instead of saying an electron, for example, is a particle just moving along, you could think that there is a wave coming in, enfolding, or it's really unfolding at a point, it's enfolded in the whole universe. Then it folds back. Then another wave comes in, at a slightly different point. You get a series of points that are very close together so we imagine they're a particle, right? But, because of that wave nature, from which it comes, you understand, the wave/particle nature, that is, it's a wave.[90]

Support for this view is found in an interpretation of holodynamics put forth by a consciousness researcher at UC Irvine, Gordon Globus, who describes Bohm's holodynamic process as follows:

> There are two primary phases of this holodynamics according to Bohm: *implication* and *explication* . . . . Implication is the "enfolding" of world to the whole and explication is the cotemporaneous "unfolding" of world from the whole. Each

[90] Bohm, *Beyond Limits*. The source is an interview with Bill Angelos; the quoted dialogue begins 32:23 into the interview. The interview runs 1:08:13 total.

> moment of unfolding/enfolding has a brief duration . . . . The world is explicate order continually unfolded from implicate order, rather than explicate order persisting.[91]

And what then might be the temporal rate of such an oscillation or "brief duration" of unfolding of the world in space–time? Here again it is the Planck constant that sets the limiting boundary condition. The Planck second, or "Planck time," calculated to be 5.44×10^{-44} seconds, is the smallest conceivable time interval consistent with quantum theory, and below this interval time has no meaning. If indeed time is granulated at such a limit, then the cosmos may be created, destroyed, and recreated a staggering number of times per "human second." In the limiting case, space–time reality can be seen to emerge from an implicate order into an explicate order, from flux into space–time, at a recurring clock cycle every 5.44×10^{-44} seconds, only to collapse out space-time and back into the implicate order in alternate intervals between each of these sub-quantum clock cycles. Such a cyclic pattern of unfolding potential accords well with Bohm's description of the holomovement here, in a 1990 paper:

> All things found in the unfolded, explicate order emerge from the holomovement in which they are enfolded as *potentialities*, and ultimately they fall back into it. They endure only for some time, and while they last, their existence is sustained in a constant process of unfoldment and re-enfoldment, which gives rise to their

[91]Globus, *The Transparent Becoming of World*, 50.

> relatively stable and independent forms in the explicate order.[92]

Meaning, Form, and Information

In developing an appreciation of the ontological and cosmological implications of quantum theory, Bohm placed great importance in maintaining a clear distinction regarding three terms: meaning, form, and information.

He describes the essential significance of form as being the conveyor and shipping container of embedded meaning:

> What is essential for a form to constitute information is that it shall have a meaning . . . . The form in the radio wave thus literally "informs" the energy in the received . . . this form is then *transformed* ("form carried across") into related forms of sound and light.[93]

How then is the meaning unpacked from the shipping container? How is information related to meaning? Bohm says that for information to become meaning requires, specifically, what Bohm terms "active information."[94]

> The basic idea of active information is that a *form*, having very little energy, enters into and directs a much greater energy. This

[92] Bohm, "A New Theory of the Relationship of Mind and Matter," 51.

[93] Bohm, "Meaning and Information," 45.

[94] Bohm and Peat, *Science, Order, and Creativity*, 84.

> notion of an original energy form acting to "inform," or put form into, a much larger energy has significant applications in many areas beyond quantum theory.[95]

For Bohm, active information is the state in which a receiving energy is informed by the *form* of a transmitted energy, the receiving energy is "in-formed." Here he gives as an example a radio wave of electromagnetic energy being "in-formed"

> In the radio wave, the form is initially inactive, but as the form enters into the electrical energy of the receiver, we may say, that the information becomes active. In general, this information is only potentially active in the radio wave, but it becomes actually active only when and where there is a receiver which can respond to it with its "own energy."[96]

Bohm goes on to say that the operation of active information can be detected in nature outside of the human context, and can be seen, for example in the action of RNA as it decodes information locked in the form of DNA:

> it is assumed that the DNA molecule constitutes a code (i.e. a language), and that the RNA molecules "read" this code, and are thus in effect "informed" as to what kind of proteins they are to make.[97]

---

[95] Ibid.

[96] Bohm, "Meaning and Information," 45.

Finally, Bohm identifies the *activity* of the energy in the receiving entity as the *meaning*, (i.e., the active information flow *is* the meaning):

> I would like to suggest then that the activity, virtual or actual, in the energy and in the soma *is* the meaning of the information, rather than to say that the information affects an entity called the mind, which in turn operates somehow on the matter of the body.[98]

What Bohm is implying here is that perhaps there is, in actuality, no "thing" that can be called mind, that there is no "entity called the mind," but that, rather, it is the *active information* itself, sourced in the implicate domain, that is in-forming the region of the explicit domain, directly. Bohm concludes:

> The thinking process unfolds certain explicate thoughts, then folds back into the implicate, and then the next thought appears, the entire process forming a series of thoughts, which, seeming not too different, appear to be flowing in a continuous stream.[99]

Active Information and the Quantum Potential

To better understand "active information," imagine that from each Planck length diameter "point"

---

[97] Ibid., 2.

[98] Ibid.

[99] Bohm, *Beyond Limits*, 5:30.

or "holosphere" in the universe (but outside of space–time) radiating waves of information encoded energy emerge from the implicate domain at the center, radiating outward into space–time in expanding shells. Active information, like conceptual thought, does not need to be of great magnitude to be the effective cause of relatively large orders of magnitude of effect in the explicate domain. An example given by Bohm is the action of a tiny electrical signal that guides the flight of a large ship, airplane, or missile in flight. Here he provides a common example of how, even in the explicate world, a source of active information, comprising relatively minute quantities of energy, can effectively influence a receiver operating at considerably higher energy levels:

> As an example, let us take a radio wave, whose form carries information representing either sound or pictures. The radio wave itself has very little energy. The receiver, however, has a much greater energy (e.g. from the power source). The structure of the radio is such that the form carried by the radio wave is imposed on the much greater energy of the receiver.[100]

Here Bohm claims that it is not the magnitude of energy that is important for active information to be effective, but *only the form itself*, guiding larger sources of energy in the explicate system. Bohm developed his mathematics of active information by deriving a potential function directly from the

[100] Ibid., 1.

Schrödinger equation in quantum mechanics, and designating it as *Q*, the "quantum potential":[101]

> As time went on, Bohm emphasized the role played by the quantum potential. The quantum potential—whose influence depends on its overall form and not on its strength—is totally unlike anything previously postulated in physics.[102]

In order to prove that there is *real* causation acting in quantum mechanics, rather than accepting the commonly held theory that it is all probability and stochastic process, Bohm proposed that a quantum potential (*Q*) also acts upon quantum particles such as electrons and protons and states that this *Q* function

> expresses the activity of a new kind of implicate order. This implicate order is immensely more subtle than that of the original field, as well as more inclusive, in the sense that not only is the actual activity of the whole field enfolded in it, but also all its potentialities, along with the principles determining which of these shall become actual.[103]

Bohm derives the quantum potential mathematically by taking derivatives of quantum mechanical Hamiltonian models of particle motion to give the following equation for the quantum potential.[104]

---

[101] Globus, *The Transparent Becoming of the World*, 55.

[102] Peat, *Infinite Potential*, 220.

[103] Bohm, "Hidden Variables and the Implicate Order," 43.

$$Q = \frac{-h^2 \nabla^2 |\psi|^2}{2m |\psi|^2}$$

Equation 1. Equation of the Quantum Potential.[105]

In Bohm's equation of the quantum potential (Q) of active information, the function ψ is the quantum field or "wavefunction" derived from Schrödinger's equation, h is Planck's constant, and m is the mass of the electron or "particle."[106] The symbol ∇ is called a "curl," and this symbol indicates an infinitesimal rotation of multidimensional coordinates (usually three or more) around a common origin, in order to obtain the differential motion of the wave in space–time. Bohm here points out what appears to be of special importance:

> What is mathematically significant in the above equation is that the wavefunction is found in both the numerator and the denominator . . . . The fact that ψ is contained both in the numerator and the denominator for Q means that Q is unchanged when ψ is multiplied by an arbitrary constant. In other words, the quantum potential Q is *independent of the strength, or intensity,* of the quantum field but *depends only on its form.* This is a particularly surprising result.[107]

---

[104] Bohm and Peat, *Science, Order, and Creativity*, 81.

[105] Ibid.

[106] Ibid.

[107] Ibid., 82. Italics added.

Of great significance for Bohm is that fact that Q, the quantum potential, is seen to be independent of strength, or signal energy magnitude. This provides support for his insistence on causality, by implying that there is a formal cause, independent of magnitude, below the postulated randomness that is otherwise implied by the probabilistic calculations of quantum mechanics, and here Bohm concludes:

> "In the causal interpretation, the electron moves under its own energy, but the information in the *form* of the quantum wave directs the energy of the electron."[108]

This can be visualized in the figure below, (Fig. 12, "Active Information and the Quantum Wave"), in which the function Q, sourced in the implicate order, molds or in-forms the quantum wave ψ in the explicate order. Here the metaphor of waves influencing the course of a ship is applicable.

---

[108] Ibid.

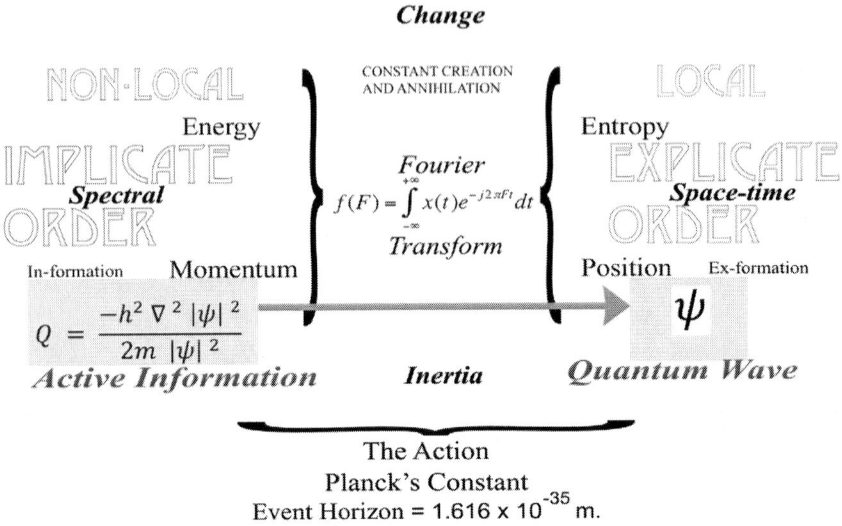

Figure *4*. Active Information and the Quantum Wave. Graphic by author.

In fact, a "pilot wave theory" similar to Bohm's active information model was proposed by de Broglie in 1927, though his idea was immediately criticized by the skeptical Wolfgang Pauli for being untestable.[109] In consequence, de Broglie abandoned his suggestion in favor of probability theory, but some years later Bohm brought up the idea again when he published, in 1952, "A Suggested Interpretation of the Quantum Theory in Terms of 'Hidden Variables.'"[110]

In the figure, the Active Information function can be seen to be guiding the Quantum Wavefunction. Bohm here provides another example of *form* as active information in everyday human experience:

---

109 Bohm and Hiley, *The Undivided Universe*, 38–39.

110 Ibid., 399.

> It is clear, of course, that the notion of active information also applies directly to human experience. For example, when the form of a road sign is apprehended in the brain and nervous system, the form is immediately active as meaning.[111]

But not all information is active, and Bohm refers to information which is not active as *virtual information*, "the activity of meaning may be only virtual, rather than actual. Virtual activity is more than a mere potentiality . . . it is a kind of suspended action."[112] And until it is made explicate as active information, this virtual information exists only in the implicate domain.

Indeed, Bohm's logic leads him to the conclusion that consciousness, at least of the nonactive kind, ("non-minding" as Karl Pribram might have said) is of, or resides "within," the implicate order (though, of course, the word "within" has no clear meaning for the spaceless, timeless implicate order.) And while he speaks of the actual "substance" of consciousness as residing primarily in the implicate order, he sees the ranges of "human experience" of consciousness residing in reality as a whole, in a continuum of the implicate/explicate orders:

> In some sense, consciousness (which we take to include thought, feeling, desire, will, etc.) is to be comprehended in terms of the implicate order, along with reality as a whole . . . the actual "substance" of

[111] Bohm, "Meaning and Information," 45.

[112] Ibid., 46.

> consciousness can be understood in terms of the notion that the implicate order is also its primary and immediate actuality.[113]

Bohm goes on to point out that some forms of information *from within* the implicate order are perceived, or can be perceived, by modes of consciousness in the explicate order: "Evidently this order is *active* in the sense that it continually flows into emotional, physical, and other responses." And here he provides as an example, music: "In listening to music, *one is therefore directly perceiving an implicate order*."[114] But it is not only in listening to music that consciousness resonates in the implicate order, but also in ordinary, everyday experience, as Bohm says, perhaps even further:

> A great deal of our difficulty comes from the fact that we accept the idea that not only matter, but all of our experience as well, is in the explicate order, and then suddenly we want to connect this up with consciousness, which is of a totally different order . . . we should consider the nature of ordinary, everyday experience. I say that it is totally misunderstood, that it is actually part of the implicate order . . . in that order everything contacts everything else and thus there is no intrinsic reason why the paranormal should be impossible.[115]

---

[113] Bohm, *Wholeness and the Implicate Order*, 196–97.

[114] Ibid., 200.

[115] Bohm, "The Physicist and the Mystic," 196.

In a 1978 interview at the Krishnamurti center in Ojai, by which time Bohm had been practicing meditation for over a decade, Bohm comments on distinctions between different modes of consciousness, and how it may be possible to experience the implicate order as a mode of consciousness which is timeless:

> **Question**: When mystics use the visualization of light they don't use it only as a metaphor, to them it seems to be a reality. Have they tapped into matter and energy at a level where time is absent?
>
> **Bohm**: It may well be. That's one way of looking at it. As I've suggested the mind has two-dimensional and three-dimensional modes of operation. It may be able to operate directly in the depths of the implicate order where this timeless state is the primary actuality. Then we could see the ordinary actuality as a secondary structure that emerges as an overtone on the primary structure . . . . The ordinary consciousness is one kind of music, and the other kind of consciousness is the other kind of music.[116]

The Sense of "I" and the Implicate Order

Bohm argues that none of this indicates that consciousness and the world of matter are absolutely separate, as assumed by Cartesian dualists. Consciousness and matter form a continuum, a whole, as Bohm says here in the final words of his last book,

---

[116] Bohm, "The Implicate Order and the Super-Implicate Order," 45–46.

The Undivided Universe, and he uses the metaphor of a mirror:

> At no stage is there a break in this process . . . . There is no need, therefore, to regard the observer as basically separate from what he sees, nor to reduce him to an epiphenomenon of the objective process. Indeed, the notion of separateness is an abstraction and an approximation valid for only certain limited purposes. More broadly one could say that through the human being, *the universe is making a mirror to observe itself* . . . . All proposals are points of departure for exploration. Eventually some of them may be developed so far that we may take them as working hypotheses.[117]

Here it may be useful to examine the conception of dualism versus monism. In Fig. 13, "Dualism vs. Monism," Cartesian dualism, denoted to the left, contrasts with three positions of Monism; Bohm's "neutral monism" can be seen in the lower right on the diagram, where matter and mind are both shown as derivative (both are in Bohm's explicate order of space–time), while the fundamental "3rd substance" would be Bohm's implicate order.

[117] Bohm, *The Undivided Universe*, 389–90. Italics added.

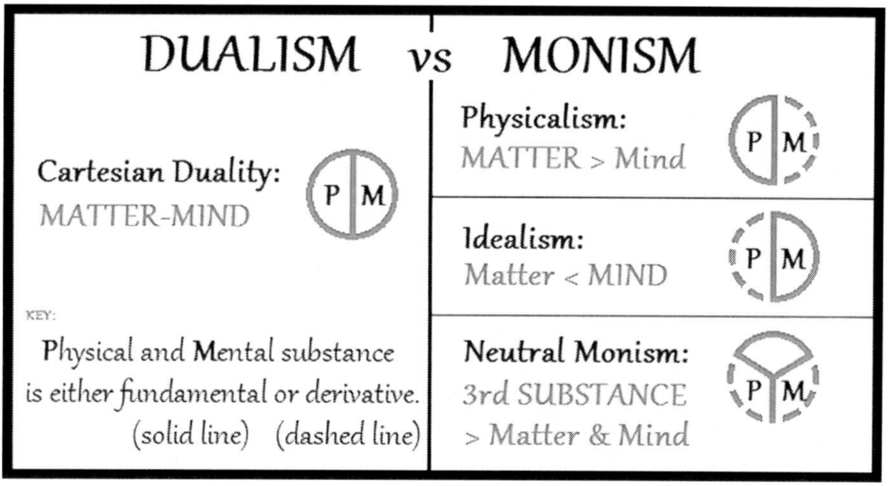

Figure 5. Dualism vs. monism. Graphic by Dewynne (2012). Reprinted under the terms of a Creative Commons CC0 1.0 Universal Public Domain Dedication. Image retrieved from Wikimedia Commons.

In an interview videotaped in Amsterdam in September of 1990, by the Dutch Public Television Network, Bohm discusses the phenomenon of the sense of Self, that sense that individuals have of being "I," and in his interview he states clearly that this sense of self is a *product* of thought, and not the real situation. Rather, he points to the implicate order as being the source of deeper reality, as being the true "I," "the Me" which we all assume to be so very unique:

> **Bohm**: Remember, thought is conditioned reflex at a very high subtle level. It just goes by itself, but it has in it the thought that thought is being produced by a center, which it calls "Me." And all the feelings, which should belong to that center, are thrown onto Consciousness as if from "Me," right?

Interviewer: And our society reinforces that.

Bohm: Everybody says the same thing, like everybody sees the same rainbow, everybody sees the same "Self," whether it's there or not.

In fact, it becomes all-important, the concept is regarded as central and supremely important. All sorts of powerful feelings will arise if it's questioned. You see, now, it all goes on by itself...and, you see... ...there's a deeper being, I say, which can...which is...that being which may be able to reveal itself anew...rather than being a fixed being...it's deep in the Implicate Order...in the Infinite.

Interviewer: So, thought creates "the Me," rather than "Me" creating the thought?

Bohm: Yes. If you don't see that happening, you will treat that as real and give it apparent reality. The whole being will take actions on that basis which seems to be coming from the center.[118]

Bohm's Quantum Theories and Consciousness

Bohm's interest, like Pribram's, also lay in trying to understand and map consciousness and its role in cosmology, an area outside of the normal range of inquiry in both their respective professional careers. Bohm was forthright in his inclusion of consciousness as an important aspect of his quantum cosmology. According to Bohm there is a two-phase reciprocal

[118] Bohm, *Beyond Limits*, 5:30.

movement or holomovement at the boundary of (between) the implicate order and the explicate order,[119] and in a 1978 interview, he states unequivocally, "Let me propose that consciousness is basically in the implicate order."[120]

In the closing words of his final published work, Bohm reifies his proposal that consciousness and quantum mechanics are connected, that "they have the implicate order in common."[121]

> Several physicists have already suggested that quantum mechanics and consciousness are closely related and that the understanding of the quantum formalism requires that ultimately we bring in consciousness in some role or other . . . the intuition that consciousness and quantum theory are in some sense related seems to be a good one . . . . Our proposal in this regard is that the basic relationship of quantum theory and consciousness is that they have the implicate order in common.[122]

Bohm's Waves Coming into and Going Out of a Point

David Bohm saw a similar process underlying what proponents of quantum mechanics theorized to

---

[119] Peat, *Infinite Potential: The Life and Times of David Bohm.*

[120] Bohm, "The Enfolding-Unfolding Universe," 62.

[121] Bohm and Hiley, *The Undivided Universe*, 381.

[122] Ibid., 381–82.

be simply "quantum jump" of energy, and here in an interview, he describes the reality as waves coming in to a point and waves coming out of a point, for all points in the holoplenum:

> **Angelos**: Could we... Again, the term "quantum" itself has its base in?
>
> **Bohm**: Well, the "quantum"... That's one of the features of "quantum"... The quantum has its base in the fact that energy is transferred in discrete jumps or "quanta" rather than continuously...
>
> **Angelos**: I see.
>
> **Bohm**: Now, you can see that some sort of quantum appears here... the wave comes to a point, then there is a jump to another point. See one thing—according to quantum theory—was that the electron could go from one state to another without passing in between. Now you said: That is an utter mystery, right? But you see, if the wave comes in to this point... then it spreads out... it could come in to another point. So, therefore, it needn't look so mysterious.[123]

To more fully grasp the relationship of consciousness to the implicate order and the explicate order, two visual metaphors are now considered: the metaphor of the micro black hole, and the metaphor of the Yin/Yang symbol.

---

[123] Bohm, *Beyond Limits*, 7:20.

Metaphor of the Quantum Black Hole

It is appropriate here to use the image of a quantum black hole as a visual metaphor for realizing the topology of the implicate order.

> Micro black holes, which have also been called quantum mechanical black holes, quantum black holes, or mini black holes, are hypothetical tiny black holes . . . . It is possible that such quantum primordial black holes were created in the high-density environment of the early Universe (or Big Bang). [124]

Visualizing the implicate order as a micro black hole, we can perhaps view its boundary as a spherical event-horizon shell, marking the limits of an "external," explicate order and the outer limits of an "internal," implicate order. Below the surface of such a sphere, of diameter equal to the Planck length, lies a transcendental implicate order, while outside the quantum black hole begins a universe of the explicate order, replete with time and space.

With this image in mind, it is appropriate here to review the topological understanding of a "point. The conventional definition of a point is a mathematical one, that it has no diameter at all, or that it is a sort of sphere with an infinitely small diameter. But a cosmological understanding of the point, considering that below the Planck length there is no space as we know it, the closest approximation to a point in space–

---

[124] Carr and Giddings, "Quantum Black Holes."

time would seem to be a micro black hole of Planck length diameter.

Metaphor of the Yin/Yang Symbol

Bohm's cotemporaneous enfolding/unfolding process can be visually captured in the Yin/Yang symbol (Fig. 14), attributed historically to Chinese Taoist sages of the third century BCE.[125]

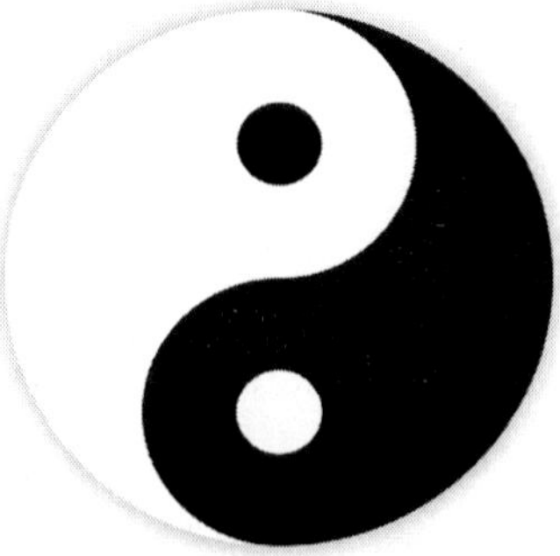

Figure 6. Implicate/explicate as yin/yang.[126] Graphic by Alkari (2013). Public domain image under the terms of Title 17, Chapter 1, Section 105 of the US Code Image retrieved from Wikimedia Commons.

In the Yin/Yang symbol, one side can be taken to represent a timeless, spaceless implicate order, and the other side can be seen as the expressed space–time explicate order. And yet each one is also at the heart of

---

[125] Wong, *The Shambhala Guide to Taoism*, 124–31.

[126] This is the Taoist yin-yang symbol, with black representing yin and white representing yang. It is a symbol that reflects the inescapably intertwined duality of all things in nature, a common theme in Taoism.

the other (designated by the dots in the two regions of the symbol).

It is of interest to note that the physicist Niels Bohr, famous for his visionary conception of electrons moving in fixed orbits around the nucleus of an atom, selected this Taoist symbol for his personal seal to adorn his "coat of arms" on the occasion of receiving the Royal Danish Award for Physics (Fig. 15).[127]

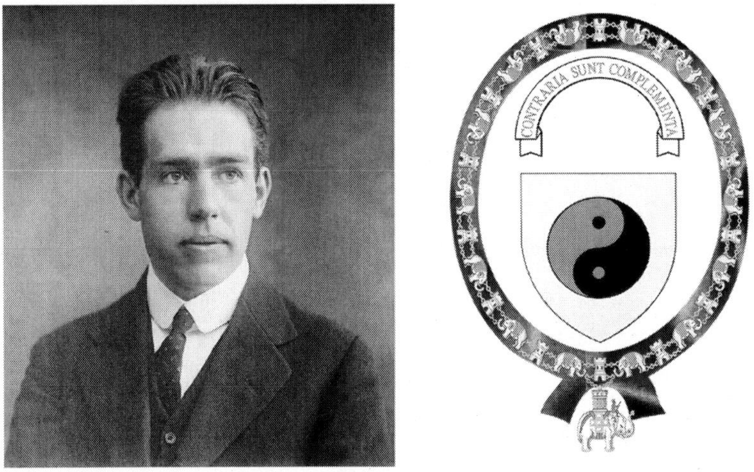

Figure 7. Niels Bohr and his coat of arms. Above left: Graphic by Bain (2012). Public domain image under the terms of Title 17, Chapter 1, Section 105 of the US Code; above right: Graphic by Fulvio314 (2013). Reprinted under the terms of a Creative Commons Attribution ShareAlike 3.0 Unported license. Image retrieved from Wikimedia Commons.

That an ancient Taoist symbol might capture so well the essence of a cosmological process that theoretical quantum physicists have only arrived at in the twentieth century speaks well for the efficacy of

[127] Globus, *Quantum Closures and Disclosures*, 175.

human consciousness in previous ages and cultures for perceiving the mechanics of the cosmos via shamanic, contemplative experience.

Chapter 3: The Holoflux Theory of Consciousness

> "Words strain,
> Crack and sometimes break, under the burden,
> Under the tension, slip, slide, perish,
> Decay with imprecision, will not stay in place,
> Will not stay still."[128]
>
> –T. S. Eliot, "Burnt Norton," (1935)

T. S. Eliot's warning here is appropriate for introducing this chapter, in which an effort is made to use words to describe a topology of an abstraction in which we attempt to integrate the theories of Bohm and Pribram. However, in *The Holographic Universe*, the writer Michael Talbot succinctly integrates the two holonomic principles:

> Considered together, Bohm and Pribram's theories provide a profound new way of looking at the world: Our brains mathematically construct objective reality by interpreting frequencies that are ultimately projections from another dimension, a deeper order of existence that is beyond both space and time: The brain is a hologram enfolded in a holographic universe.[129]

---

[128] Eliot, *Collected Poems 1909—1962*, 176.

[129] Talbot, *The Holographic Universe*, 54.

That David Bohm and Karl Pribram influenced one another and became both professional and personal friends has been noted. In Fig. 16 can be seen the two enjoying themselves in Prague, with Bohm's wife, Saral, just four months before Bohm's death in October 1992.

In this chapter the ontological platform begun with the Pribram/Bohm holonomic theories is extended through integration with properties of the electromagnetic field implicated by Pribram's approach to holonomic brain theory.

Figure *8*. David Bohm and Karl Pribram in Prague, June 1992.[130] Photo by Katherine Neville. Reprinted with permission.

The holoflux field theory of consciousness is described here as a process of an energy of awareness

130 Neville, "Saral and David Bohm," picture 8.

transforming in and out of two domains: the explicate and the implicate.

The theory is supported by a set of concepts from physics, electrical engineering, and mathematics. These concepts (bulleted below) are developed in this chapter through use of verbal description, analogy, and topological diagrams, with minimal recourse to equations or mathematical notation. In each case, focus is upon the ontological meaning of each concept:

- the Planck constant
- the holoplenum
- Wheeler's "Self-observing U"
- frequencies
- wavelengths
- Wien's law
- information
- superposition
- resonance
- dimensional analysis

The Planck Constant ($\hbar$)

Bohm makes frequent reference to a fundamental numerical value, the Planck length of 10^{-33} cm, widely used in quantum mechanical calculations, and derived from "the Planck constant" (denoted in calculations as $\hbar$). In speaking on the implicate order, Bohm states:

> It must be remembered that even this vast sea of cosmic energy takes into account only what happens on a scale larger than the critical length of 10^{-33} cm . . . . But this length is only a certain kind of limit on the applicability of ordinary notions of space and time. To suppose that there is nothing beyond this limit at all would indeed be quite arbitrary. Rather, it is very probable that beyond it lies a further domain.[131]

In 1900 the Berlin physicist Max Planck (Fig. 17), like Bohm decades later, found himself dissatisfied with what he saw as a problem with the tendency toward over-reliance on mathematical statistics, rather than the attempt to discern some mechanism which would explain more clearly the underlying reality of the phenomenon of energy and thermodynamics.

---

[131] Bohm, *Wholeness and the Implicate Order*, 193.

Figure 9. Portrait of Max Planck, 1878. Graphic by Jbarta (2014). This work is in the public domain under the terms of Title 17, Chapter 1, Section 105 of the US Code.

In his 1879 doctoral thesis, Planck discussed two major approaches to the second law of thermodynamics that were at the forefront of the physics of his day: one theory stressed the use of mathematically continuous equations to model unbroken flows of radiant field energy, while the other theory, the "atomist" theory, held that since matter is not continuous, it is preferable to apply statistics to calculate the mechanical motions of atoms and molecules in thermodynamics processes.[132] Planck felt that the atomist approach, relying as it did on statistical calculations, was masking the underlying

[132] Baggott, *The Quantum Story*, 7–8.

realities. Over the next 20 years his research focused on the search for a nonstatistical mathematical model to reflect the ontological realities that linked radiation and the phenomenon of "heat."

On December 14, 1900, a date "widely acknowledged to be the date on which the quantum revolution began," Planck presented his derivation of a new radiation equation to a meeting of the German Physical Society, and in his presentation stated, "We therefore regard—and this is the most essential point of the entire calculation—energy to be composed of a very definite number of equal finite packages."[133] He went on to describe the constant in the equation which he had derived (later to be called the "Planck constant"), and stated that, "since it has the dimensions of a product of energy and time, I called it the elementary quantum of action." Planck postulated that electromagnetic energy both absorbed and emitted only in these elementary quanta of actions, or discrete packets of energy, which he also referred to as "energy elements." Planck's new theory, and his constant, published a month later in a 1901 issue of the journal *Annalen der Physik*, soon became a cornerstone of quantum mechanics.[134]

Earlier, in 1893, William Wien had discovered a mathematical relationship that allowed him to calculate the peak frequency of electromagnetic radiation for a given temperature, but his equations were only accurate for extremely high temperatures. Planck's paper solved the problem by assuming that

[133] Ibid., 15.

[134] Ibid., 15.

any measurable electromagnetic fluctuation for a given temperature must consist of a unique spectrum of discrete frequencies, an integral set of separately calculable single frequency oscillators, each one making its own radiant contribution to the whole. Using his derived constant, Planck was able to calculate precisely the energy, ***E***, radiating in all directions from a theoretical "point" for a given macro "temperature."[135] The accepted value of this constant, as determined by Planck, is $\hbar = 6.62606957 \times 10^{-34}$ Joule-seconds.[136]

The Planck constant was derived specifically from other fundamentally agreed upon physical constants, including gravity and the speed of light. In presenting his constant, Planck expressed the importance of this constant in rather mystical terms, claiming that the discovery paved the way for, as he described:

> The possibility of establishing units of length, mass, time and temperature which are independent of specific bodies or materials and which necessarily maintain their meaning for all time and for all civilizations, even those which are extraterrestrial and nonhuman, constants which therefore can be called with surety, "the fundamental physical units of measurement."[137]

---

[135] Planck, "On the Law of Distribution of Energy in the Normal Spectrum."

[136] Ibid.

[137] Baggott, *The Quantum Story*, 11.

In his 1901 paper, Planck presented a simple equation to describe the energy output of a single "energy element" or "quantum." The equation is:

$$\varepsilon = h\nu$$

In this equation the first term, ε, the energy element, is equal to the product of the constant (Planck's constant, h) times the frequency, ν. The use of the constant h to unify, or normalize, the basic cosmological constants in physical calculations was immediately found to be so useful to simplify previously complex calculations, that the number was soon referred to as "Planck's constant," although in reality his discovery involved a set of constants, several of which are listed here in Table 2.[138]

---

[138] Benenson, Harris, Stocker, and Holger, *Handbook of Physics.*

Table 1
Several Plank Constants

Constant	Value
Length	1.616199×10^{-35} meter
Time	5.39106×10^{-44} seconds
Area	2.6121×10^{-70} square meter
Power	3.62831×10^{52} watts

Note. Data adapted from Benenson et al. (2006), *Handbook of Physics*, 47. Author's table.

In the spring of 1905, a 26-year-old named Albert Einstein published four papers, soon to become famous in the world of physics. His first paper extended Planck's theory to include light, by asserting that Planck's radiation law only made sense if radiant energy of all wavelengths consists of discrete packets of energy, which Einstein called "light-quanta," or "photons."[139]

To obtain a better feel for the magnitude and use of these constants, let us calculate the velocity of one photon as it travels the distance of one Planck length in one Planck second. The equation for velocity is: velocity = length / time. If we substitute the Planck values for length and time (values found in Table 2), the velocity of a photon at the Planck level can be calculated as follows:

[139] Baggott, *The Quantum Story*, 21.

The equation: *Velocity* = Planck Length/Planck Time

The values used: Planck Length = 1.616199×10^{-35} meter

Planck Time = 5.391060×10^{-44} seconds

Substituting: ***Velocity*** = 1.616199×10^{-35} meter / 5.39106×10^{-44} seconds

Results in the following velocity of the photon:

Velocity = 2.99792×10^{8} meters / second

This happens to be the exact speed of light in a vacuum.

If we try to move faster by decreasing the time it takes to move the same distance, the denominator in the velocity equation would have to be a smaller value, and a smaller value in the denominator would make the equation produce a higher velocity, violating the ceiling to the speed of light in a vacuum. Conversely, if we keep the same time interval and try to increase the distance travelled to something greater than the Planck length, we would again be increasing the velocity, exceeding the speed of light. Accordingly, there can exist no measurable (or conceivable) space smaller than that of the Planck length, and there can be no time interval smaller than that of the Planck second.

The Planck "second," then, at roughly 10^{-44} human "seconds," is the smallest possible time measurement, according to quantum mechanics, and is understood as a time period below which time has no meaning.[140] According to contemporary models of cosmological origin, this number also happens to be the time marked as the very beginning of our universe, at which time expansion into space began:

> The early inflationary universe is proposed to have had an accelerated expansion, the duration of which depends on the particular inflationary model. In any case, inflation is believed to have lasted an extremely short time: in the range of 10^{-33}–10^{-35} sec. In particular, in the chaotic inflationary model, inflation starts at the Planck time, $t_P = 10^{-44}$ sec.[141]

Similarly, the Planck length (at 10^{-35} meters, or its equivalent, 10^{-33} cm) is the smallest measurable length in physically defined space, below which space as we know it has no meaning. Below the Planck length one cannot say "smaller," because there is no spatial dimension definable below 10^{-35} meters. It is worth considering the contrast, between the Planck length of at 10^{-35} meters, and the diameter of a proton at 10^{-15} meters. The proton is enormous at an order of 20 magnitudes larger than a sphere of Planck length diameter.

It is also of interest to examine the last entry in the list of Planck constants shown in Table 2, the "Power" value, which indicates that a photon vibrating at the dimensional frequency corresponding to one

[140] Wheeler, *A Journey into Gravity and Spacetime.*

[141] Zizzi, "Consciousness and Logic in a Quantum Computing Universe," 459–60.

Planck wavelength contains unimaginable power. The raw power flowing out from all of the stars in the known universe has been estimated to be 2×10^{49} watts.[142] The Planck Power, by contrast, has been calculated to be an incredible 10^{52} watts, three orders of magnitude *greater* than the luminosity of all of the stars in the visible universe combined.[143]

While quantum physics indicates that below the limiting Planck length there can be no space–time as we know it, Bohm predicts that it is precisely there, below the inner space–time boundary, that manifests a nonspatial, nontemporal domain which he identifies as "the implicate order."[144] The relationship of explicate order to implicate order should not necessarily be understood as a one-to-one relationship. On the contrary, as Bohm conjectures here, in a 1987 paper, there may in fact be *multiple explicate orders* (supporting the widely held "multiverse hypothesis"), suborders of a single, omni-connected implicate order:

> The principles of organization of such an implicate order can even define a unique explicate order, as a particular and distinguished sub-order, in which all the elements are relatively independent and externally related. To put it differently, the explicate order itself may be obtainable from the implicate order as a special and

---

[142] Ledrew, "The Real Starry Sky."

[143] Benenson, Harris, Stocker, and Holger, *Handbook of Physics*.

[144] Bohm, *Wholeness and the Implicate Order*, 190.

determinate sub-order that is contained within it.[145]

The Sub-Quantum Holosphere

We have seen that Bohm proposes two orders, an explicate (space–time) order and what he terms "an implicate order beyond the critical limit of 10^{-33} cm," a sub-quantum (below space–time) order.[146] Having established the relative location of these two domains, it is useful to further identify the structural relationship of the implicate order to the explicate order by considering another image. The previous metaphor (Fig. 9) was presented in the image of an "ocean depth," visualized in a vertical "up/down" geometry, having a "sub-quantum bottom" at the Planck depth of 10^{-33} cm. This ocean depth metaphor can be improved with a new metaphor based upon cosmology, that of a three-dimensional "outer/inner" shell of spherical geometry.

In the following diagram (Fig. 18, "A Sub-Quantum Holosphere"), the implicate order is seen as having its origin at the center of space, beginning at a spherical shell with a Planck length diameter of 10^{-33} cm, manifesting a spherical region that the physicist Paola Zizzi calls a "fuzzy quantum black hole."[147] According to Zizzi, "its event horizon is a fuzzy sphere."[148]

---

[145] Bohm, "Hidden Variables and the Implicate Order," 44.

[146] Ibid., 100.

[147] Zizzi, "Consciousness and Logic in a Quantum Computing Universe," 469.

Here, within the deepest "inside" possible in a space–time universe and located at the geometric center of every point in the space–time universe, lies the implicate order. In the figure, the sub-quantum holosphere or implicate order is depicted at the center of a spherical explicate region of the space–time universe, at the boundary separating time from the nontemporal. The holosphere is fixed and does not move, while around it move waves of space–time energy. The previous scale from Fig. 9 is shown here as a radius extending outward from the implicate order within the central sub-quantum holosphere.

---

[148] Ibid., 468.

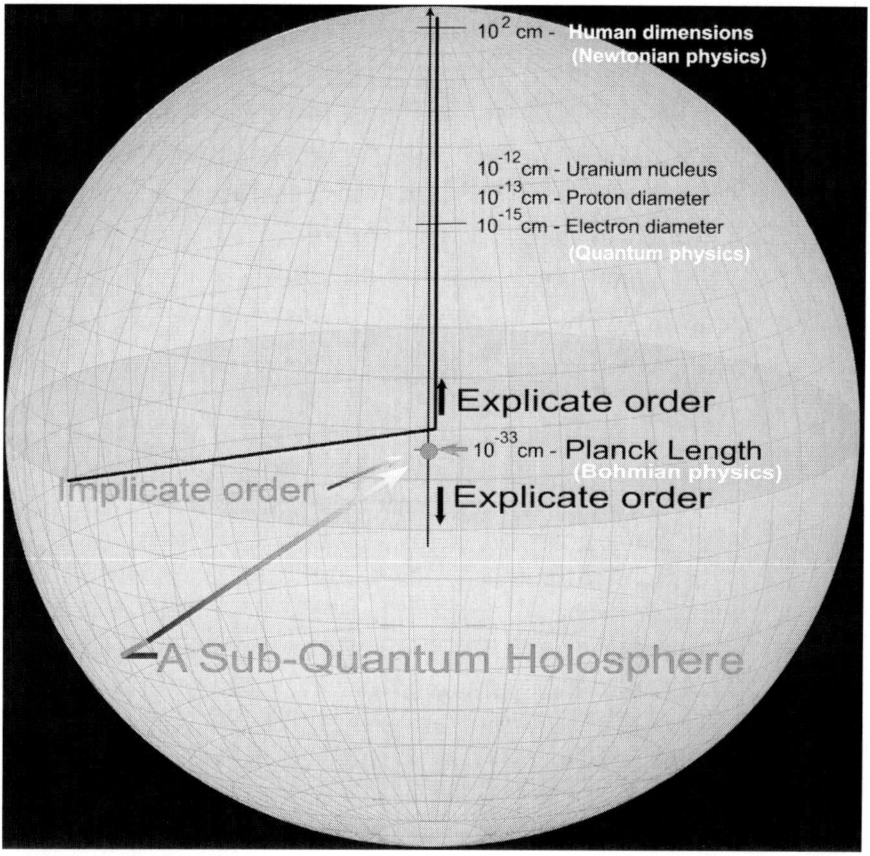

Figure 10. A sub-quantum holosphere. Dimensional data from Benenson et al. (2006), Handbook of Physics. Graphic by author.

Again, the words of the poet T. S. Eliot may here be appropriate:

> And do not call it fixity,
> Where past and future are gathered. Neither movement from nor towards,
> Neither ascent nor decline. Except for the point, the still point,
> There would be no dance, and there is only the dance.[149]

The Sub-Quantum Holoplenum

This concept of a sub-quantum holosphere is mirrored both in Bohm's conceptualization of the implicate order as well as in contemporary theories of quantum computing, where quantum black holes are seen to carry and store information in the event horizon, described here by P. Zizzi, a physicist at the University of Pavia, Italy:

> The event horizon of such a quantum black hole is the surface of a fuzzy sphere . . . . Fuzzy quantum black holes and quantum human minds can be viewed as subroutines of the whole quantum-computing universe . . . . This cosmic network exploits quantum communication complexity, and, thanks to entanglement, allows its parts to accomplish a distributed task without the need of any kind of communication. This is possible because all its parts share a common quantum space–time background, which is nonlocal.[150]

The physicist John Wheeler (1911–2008) has hypothesized that a considerable amount of information is storable on the surface boundary of a spherical black hole. His theory is discussed in detail elsewhere in this chapter, but here it is worth mentioning that his calculations indicate that the surface boundary of a one-centimeter diameter spherical black hole is capable of storing 10^{64} bits of

[149] Eliot, *Four Quartets*, 16.

[150] Zizzi, "Consciousness and Logic in a Quantum Computing Universe," 468–69.

information.[151] Compared even to the total number of stars in the universe, which has been estimated to be 10^{24}, this is a great deal of potentially storable information.[152]

Assuming the existence of a sub-quantum holosphere, the next question to ask would be how many might there be, underlying our universe? Perhaps to coincide with Big Bang theory there was only one, out of which the universe of space–time emerged. An alternate approach would be to imagine the universe as consisting of an infinity of quantum black holes, each one having a diameter of the Planck length. In such a model, the implicate order would be distributed evenly, to be found at the center, everywhere, throughout the space–time cosmos. Bohm describes such a model, arguing here for a cosmological bedrock of actuality, which he calls "the plenum."[153]

> It is being suggested here, then, that what we perceive through the senses as empty space is actually the plenum, which is the ground for the existence of everything, including ourselves. The things that appear to our senses are derivative forms and their true meaning can be seen only when we consider the plenum, in which they are generated and sustained, and into which they must ultimately vanish.[154]

---

[151] Wheeler, *A Journey into Gravity and Spacetime*, 222.

[152] Van Dokkum and Conroy, "A Substantial Population of Low-Mass Stars."

[153] Ibid.

Bohm points out that in early Greece, the philosophers Parmenides and Zeno also held that "space is a plenum."[155] Here, located in a communion of Planck holosphere, we find a plausible explanation for the peculiar observed phenomenon that the universe, from every point, appears to be expanding outwardly, directed radially *away from each point*; thus, in the space–time universe, the center is everywhere:

> Relativity assumes that the universe will appear much the same no matter where the observer happens to be situated. The nebulae appear to be rushing away from our particular position, and they will appear to be rushing away from any other position in which an observer is located.[156]

Thus we hypothesize this Bohmian holospheric plenum, like some sea-floor bed of existence, to be discoverable at the center of every conceivable coordinate in space–time; the concept can be seen resonating vibrantly in the title of Jean Gebser's magnum opus, *The Ever-Present Origin.*[157] We can

---

[154] Bohm, *Wholeness and the Implicate Order*, 191–92.

[155] Ibid., 191.

[156] Kuehn, *A Student's Guide Through the Great Physics Texts: Vol 1*, 332.

[157] Gebser, *The Ever-Present Origin: Part One: Foundations of the Aperspectival World.*

find the same image resonating in T. S. Eliot's "still point of the turning world."[158]

[158] Eliot, *Four Quartets*, 15.

O hidden under the dove's wing, hidden in the turtle's breast,
Under the palm tree at noon, under the running water
At the still point of the turning world. O hidden.[159]

–T. S. Eliot, "Coriolan," (1931)

We also hypothesize that this center of the cosmos, the origin, located everywhere throughout the cosmos. Having established the ontological basis of a plenum, Bohm goes further to envision what he terms "the holomovement," the dynamics of energy processing within this plenum, an "immense sea of energy," flowing within and between the non-time-space domains of the implicate and the explicate orders:

> This sea is to be understood in terms of a multidimensional implicate order, while the universe of matter as we generally observe it is to be treated as a comparatively small pattern of excitation. This excitation pattern is relatively autonomous and gives rise to approximately recurrent, stable and separable projections into a three-dimensional explicate order of manifestation, which is more or less equivalent to that of space as we commonly experience it.[160]

This region of implicate order, the holoplenum (Fig. 19), is here visualized as consisting of close-packed holospheres, each a sphere bounded by the Planck limit of 10^{-33} cm.

[159] Eliot, *Collected Poems 1909–1962,* 126.

[160] Bohm, *Wholeness and the Implicate Order*, 192.

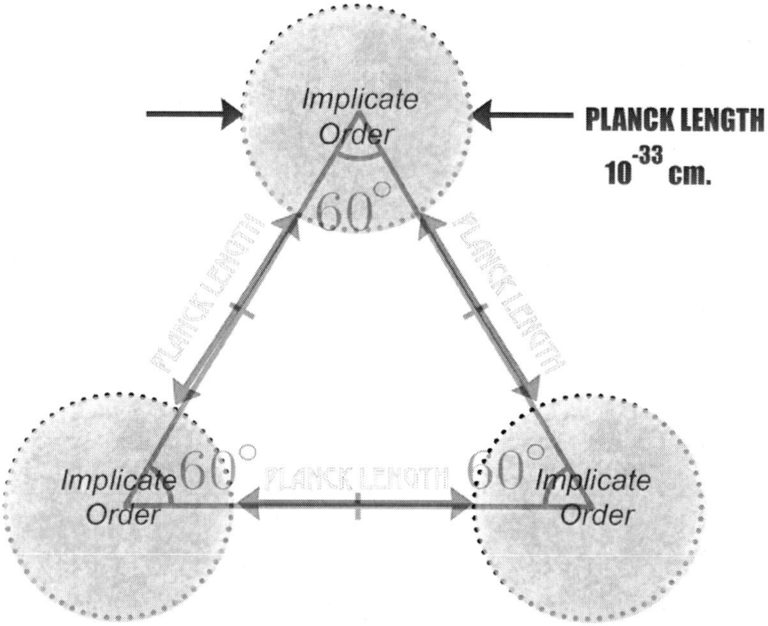

Figure *11*. Holoplenum formed by implicate order. Graphic by author.

Because each holosphere can come no closer than one Planck length to neighboring holospheres, simple geometry insures that in a single plane these holospheres form equilateral triangles, and in three dimensions the holospheres of the holoplenum would thus form regular tetrahedrons in close-packed configuration (Fig. 20). This geometry aligns with the prediction of the architect and systems theorist Buckminster Fuller, who wrote that the natural analytic geometry of the universe is based on arrays of tetrahedra; Fuller supported his conjecture with the fact that the strongest possible homogeneous truss is cyclically tetrahedral.[161]

[161] Edmondson, *The Synergetic Geometry of R. Buckminster Fuller*, 19.

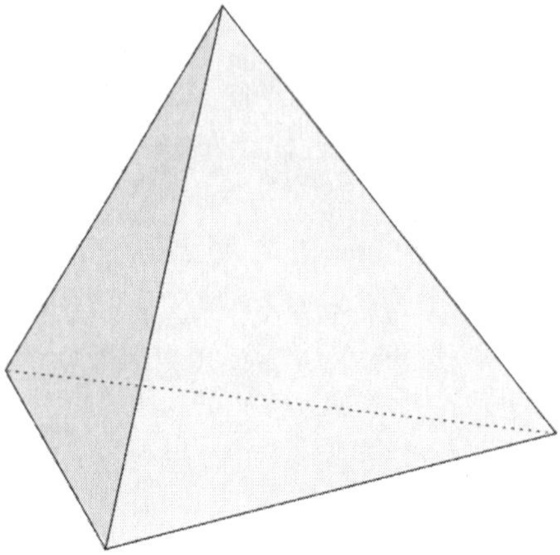

Figure 12. A regular three-dimensional tetrahedron. Graphic by author.

Fuller's idea, that a three-dimensional tetrahedron may be a natural underlying geometry of nature, has been reinforced by the discovery, in 2013, that DNA strands at the nanostructure level form stable three-dimensional tetrahedrons and therefore are suitably scalable for nanotechnology engineering:

> The physicochemical stability of DNA makes it a desirable candidate for the design of nanostructure materials as DNA is inherently nanoscalable . . . . The DNA nanostructure consists of six single-stranded oligonucleotides that hybridize to form a *three-dimensional tetrahedron.*[162]

In addition to operating within a geometrically maximally stable plenum, any phenomena at the Planck dimension would also operate at the highest

[162] Rendek, Fromme, Grotjohann, and Fromme, "Three-Dimensional DNA Nanostructure," 141.

frequency rate possible in space–time, simply because at this smallest dimension would also be found the smallest wavelength, both being equal to one Planck length. According to Wien's law (the higher the frequency, the higher the energy), at the smallest wavelength possible, at the Planck length, would also be found the highest energy.[163]

Wheeler's Self-Observing U

"Truth is less than truth until it is made known."[164]
–John Archibald Wheeler

In the context of consciousness, another visual metaphor is revealing, congruent with Bohm's cosmology. The metaphor can be seen in a relational diagram drawn by the Princeton physicist John Archibald Wheeler (1911–2008). Wheeler had worked closely with Bohr in the late 1930s, was a Princeton colleague of Bohm's in the late 1940s, and thesis advisor to a young Richard Feynman, who received the Nobel Prize in Physics in 1965.[165]

Wheeler used a visual metaphor (Fig. 21), which he termed the "Self-Observing U," to help understand the quantum theoretical observation and creation of an actual entity, within an explicate (space–time) universe. As Wheeler explains here in his 1998 memoir, *Geons, Black Holes, and Quantum Foam*, he views the universe as a self-excited circuit:

---

[163] Bohm, *Quantum Theory*, 6.

[164] Wheeler, *A Journey into Gravity and Spacetime*, v.

[165] Baggott, *The Quantum Story*, 242.

Not all potentiality is converted to actuality in any finite time . . . . My diagram of a big U (for universe) attempts to illustrate this idea. The upper left end of the U represents the Big Bang, when it all started. Moving down along the thin left leg and up along the thick right leg of the U symbolically traces the evolution of the universe, from small to large—time enough for life and mind to develop. At the upper right of the U sits, finally, the eye of the observer. By looking back, by observing what happened in the earliest days of the universe, we give reality to those days. The eye could as well be a piece of mica. It need not be part of an intelligent being. The point is that the universe is a grand synthesis, putting itself together all the time as a whole.[166]

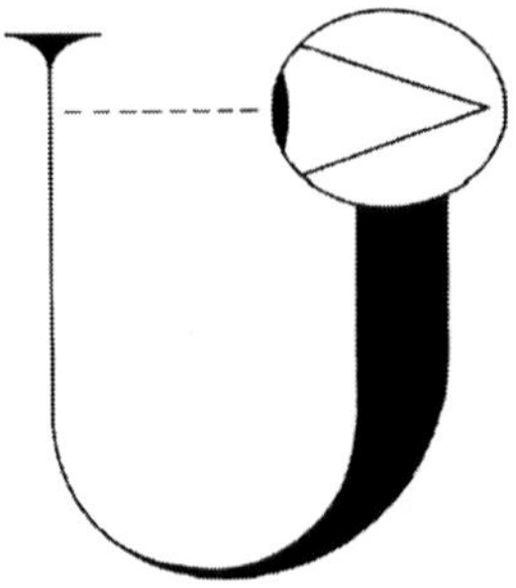

Figure *13*. Wheeler's self-observing U. Graphic by author.[167]

[166] Wheeler, *Geons, Black Holes, and Quantum Foam*, 338.

[167] Modified from figure captioned, "Does looking back 'now' give reality to what happened 'then'?"; see Wheeler, *Geons, Black Holes, and Quantum Foam*, 339.

The figure also supports Bohm's conjecture that there is no separation between the observing instrument and what is observed, as he elaborates here:

> For example, you are looking at this table; the form of this table has been built up by your experience which you are projecting into the table. Is the table you, or is it something separate from you? You appreciate it as a table with a certain form and a certain subsistence, but that form and that subsistence are as much you as the table. If you probed it with a very high energy machine, say with neutrinos, they would go right through, and the table would be a vaporous nebula. So the form of the table as a solid substance, or subsistence, comes from the human brain with its own particular mode of interacting. In a sense, the observer is the observed.[168]

That this paragraph on perception sounds a great deal like passages from Karl Pribram's 1972 *Languages of the Brain* is likely no coincidence, as Bohm had previously met Pribram at a Krishnamurti conference in Ojai, and the two had become personal acquaintances and shared many ideas over the preceding five years.[169]

---

168 Bohm, *A Question of Physics: Conversations in Physics and Biology*, 132.

169 Pribram, *The Form Within.*

Creation and Dissolution of the Universe

In the same sense that a digital computer system has a clock-cycle to which all events are synchronized, what might be the underlying clock-cycle of the universe? According to quantum theory, the clock-cycle can be no less than the Planck second, for, as previously discussed, any time unit smaller than Planck time has no meaning. As Bohm's colleague, the physicist Hiley states:

> The explicate "moments" that unfold have a certain temporal thickness, a certain duration. Moments are temporal segmentations of unfoldment . . . . This is "a moment where *what has been* is separated from *what is yet to come*." Within this brief duration time becomes nonlocal.[170]

Thus there is separation, and a cycling between local and nonlocal time, and this can be no less than the Planck length. Figure 22 illustrates such a hypothetical clock-cycle with a Planck-limited value of 5.39106×10^{-44} seconds as the clock cycles the cosmos on and off, in and out, forming the digital holoflux.

---

[170] Globus, *The Transparent Becoming of World*, 132.

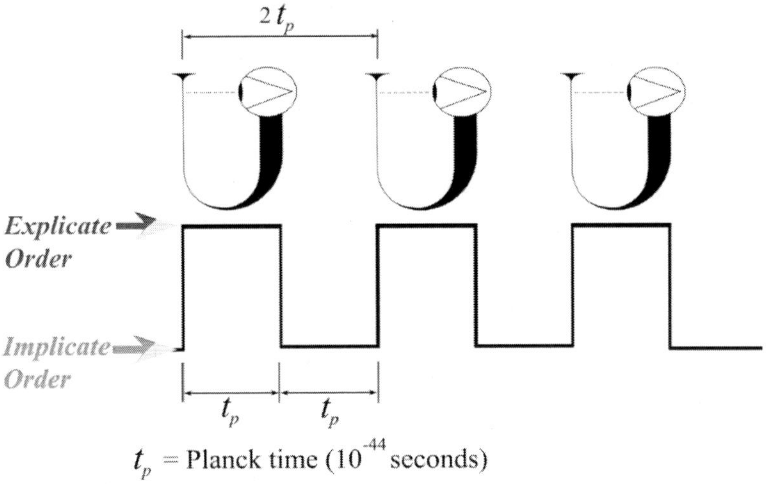

Figure 14. Clock-cycle: digital holoflux. Graphic by author.

This theoretical explicate/implicate digital clock cycle is supported in quantum theory by the concept of "quantum fluctuations," described here by Wheeler:

> Space—pure, empty, energy-free space—all the time and everywhere experiences so-called quantum fluctuations at a fantastically small scale of time, of the order of 10^{-44} second and less. During these quantum fluctuations, pairs of particles appear for an instant from the emptiness of space—perhaps an electron and an antielectron pair or a proton and an antiproton pair. Particle–antiparticle pairs are in effect all the time and everywhere being created and destroyed. Their destruction is so rapid that the particles never come into evidence at any everyday scale of observation.[171]

[171] Wheeler, *A Journey into Gravity and Spacetime*, 223.

This accords well with Bohm's ontological model, in that each quantum tick reveals a new explicate order:

> As with consciousness, each moment has a certain explicate order, and in addition it enfolds all the others, though in its own way. So the relationship of each moment in the whole to all the others is implied by its total content: the way in which it "holds" all the others enfolded within it.[172]

This holoflux clock-cycle process or oscillation between an implicate order and an explicate order is also remarkably congruent with Dewey's previously-mentioned theory of laminated space–time.[173] Going further, in considering that, within an explicate space–time order, any frequencies in time imply the existence of corresponding frequencies in space, Bohm's comments:

> These ideas are connected with our ontological interpretation by means of a model of a particle as a sequence of incoming and outgoing waves, with *successive waves very close to each other.*[174]

These "very close" waves find their successive limit, as we have discussed, at the bounding Planck length of 10^{-33} cm, and the bounding Planck time value

[172] Bohm, *Wholeness and the Implicate Order*, 207.

[173] Dewey, *Theory of Laminated Spacetime.*

[174] Bohm and Hiley, *Ontological Interpretation of Quantum Theory*, 9–10.

of 10^{-44} sec time can also be viewed as a total of 10^{44} successive alterations ("incoming and outgoing waves") per (human) second.

Temperature/Energy of a Quantum Black Hole

Looking at a calculation of the theoretical temperature of an "object" of the size of holosphere which is assumed, for purposes of calculation, to be a black hole, we can use an equation derived by Stephen Hawking to calculate the temperature at the surface of such a quantum black hole (Equation 2, "Hawking's temperature of a black hole"):

> Notice that Hawking's formula, the mass of the black hole is in the denominator. That means that the larger the mass, the colder the black hole, and conversely, the smaller the mass, the warmer the black hole.[175]

$$T = \frac{1 \quad c^3 h}{16\pi^2 GMk}$$

Equation 2. Hawking's temperature of a black hole.[176]

According to the Hawking equation and Wien's law, we see that the highest temperature, the greatest energy and highest frequency of electromagnetic energy possible, all issue from the very heart of each and every "point" in space–time, from within the theoretical locus of a holosphere of diameter 10^{-35} m, the Planck length. Thus the quantum wave, or Planck

[175] Susskind, *The Black Hole War*, 173.

[176] Ibid.

wave of quantum number one, enters the space–time continuum at maximum temperature, from every holospheric point in the holoplenum, at a cyclic frequency of 10^{44} Hz, a frequency limit set by the Planck time of 5.39106×10^{-44} seconds.

Wheeler's "It from Bit" Black Hole

Even beyond timing considerations, Wheeler also developed a theory of information storage, in which he finds at the upper limit the Planck length, deep within the physical world:

> "It from bit" symbolizes the idea that every item of the physical world has at bottom, at a very deep bottom, in most instances an immaterial source and explanation.[177]

Wheeler's explanation was to digitize the cosmos by assigning one "bit" of information to the smallest physical area possible in space–time: an area equal to that of a square having each side the Planck's limiting length of 10^{-35} meters.[178]

In 1970 one of Wheeler's graduate students, Jacob Bekenstein, proposed a novel idea, that there is a maximum amount of information that can be stored in a finite region of space, and that quantum theory points to that area:[179]

---

[177] Wheeler, *Information, Physics, Quantum: The Search for Links*, 361.

[178] Wheeler, *Geons, Black Holes & Quantum Foam: A Life in Physics*, 297.

[179] Wheeler, *A Journey into Gravity and Spacetime*, 222.

> One unit of entropy (information), one unit of randomness, one unit of disorder, Bekenstein explained to me, must be associated with a bit of area of this order of magnitude . . . . Thus one unit of entropy is associated with each 1.04 x 10^{-69} square meters of the horizon of a black hole.[180]

This proposed upper limit to the information that can be contained upon the surface of a specific, finite volume of space has come to be known as the "Bekenstein bound."[181] Figure 23 is Wheeler's depiction of information bits stored on the surface of a sphere.

---

[180] Ibid.

[181] Bekenstein, "Black Holes and Entropy."

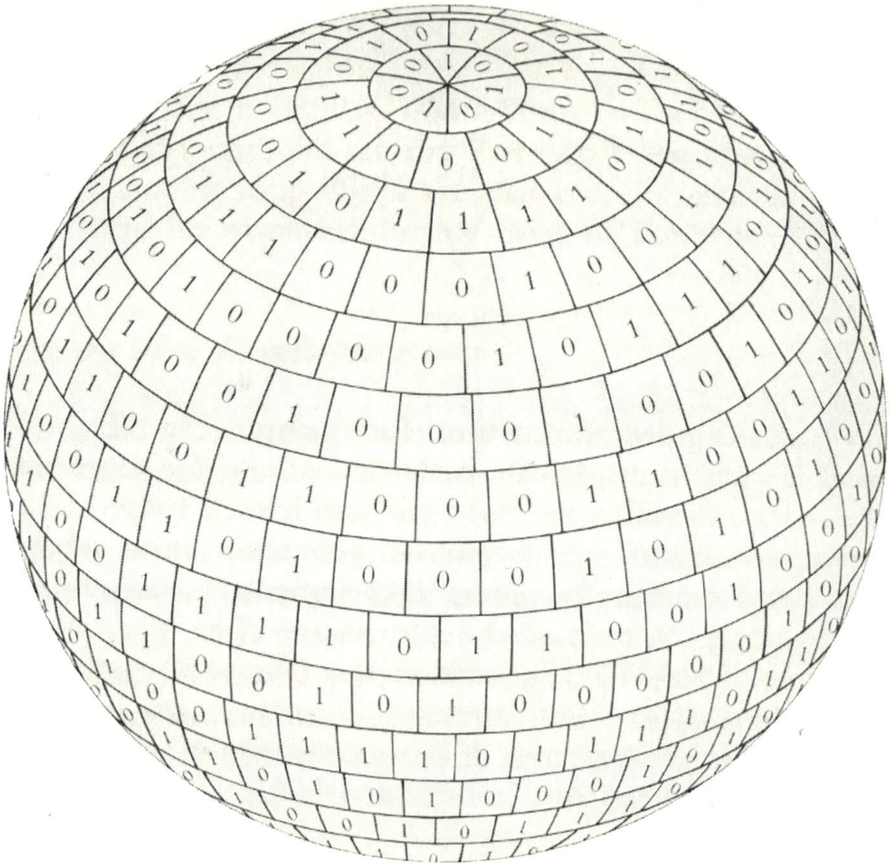

Figure 15. Symbolic representation of the Bekenstein number of bits. Wheeler (1990), A Journey into Gravity and Spacetime, 220; reprinted with permission.

Using this concept, we can calculate the maximum amount of information that might be stored in a one-centimeter radius sphere located in the human brain. To obtain this limiting number of bits, we simply calculate the surface area on a sphere of one-centimeter radius, and then divide the result by 1.04×10^{-69} square meters (a square of Planck length). Using simple geometry, we use the equation for the surface area of a sphere $4\pi r^2$. We can then approximate

the area to be, $4\pi r^2 = 4 \times 3.14 \times (10^{-2})^2$ or 1.256×10^{-5} square meters. If we divide this value by 1.04×10^{-69} square meters we obtain the following answer for the number of bits of information stored in a one-centimeter radius sphere:
$1.256 \times 10^{-5} / 1.04 \times 10^{-69} = 1.207 \times 10^{64}$ bits of information.

By contrast, if we perform the same calculation using the surface area of a holosphere (quantum black hole), the surprising answer (perhaps not so surprising) is that there are π qubits per holospherical surface, which implies that the universe, in terms of information computation, is not based on two-state binary computing, but that it operates on π-based computational algorithms at the base level of the holoplenum.

The enormous amount of information that might be physically stored in the holoplenum accords well with the holographic principle described by the theoretical physicist Paola Zizzi, the proponent of a quantum gravity theory in which space is considered to be granular. According to Zizzi, the universe should be regarded as a kind of super computer:

> The holographic principle states that all the information enclosed in a region of space with volume ***V***, is encoded on the surface ***S***, bounding ***V***. More precisely, each pixel of area (a pixel here is one unit of the Planck area) of ***S*** encodes one bit of information. In a more sophisticated (quantum) version of the holographic principle the encoded information is quantum, and each pixel encodes one quantum bit (or "qubit") of information . . . . At this point every quantized horizon can be viewed as a

quantum register (a quantum register is the memory of a quantum computer, built with qubits).[182]

The problem remains, however, *how* this information might be stored, and more importantly, in order to address Chalmers' "hard problem," what is the mechanism for comprehension of this information? How is such information experienced? To answer "how" for either of these problems first requires a clear understanding of how information is encoded, transmitted, and received via twenty-first-century human technology. We begin with the concepts of frequency and the frequency domain.

Fields, Frequency, and the Frequency Domain

An understanding of fields and frequency, their measurement and dynamics, are of central importance to an understanding of the holoflux field theory of consciousness. It was Michael Faraday (1791–1867) who first conceived of the reality of an invisible electromagnetic "field" during his observation of magnetic lines of force around a conductor carrying a direct current.[183] But while Faraday had the ability to conceptualize a certain ontological reality, he did not have sufficient mathematical training to formalize the dynamics of his theorized field. Thus it was James Clerk Maxwell (1831–1879) who in 1861, through a series of interlocking equations, subsequently known

[182] Zizzi, "Consciousness and Logic in a Quantum Computing Universe," 460.

[183] Crease, *The Great Equations*, 135.

as the "Maxwell equations," was able to model the dynamics of electric and magnetic energy, in waves of various frequencies, flowing through space–time in an ocean of vibrating flux.[184]

Figure 16. James Clerk Maxwell at Trinity College, Cambridge. Graphic by Tjlaxs (2005). Public domain image published in the US before 1923. Retrieved from Wikimedia Commons.

His discovery was a mathematical one, based upon his development of a set of four interlocking partial differential equations with which he was able to predict the dynamics of electric charges and how they are generated by oscillating magnetic fields. Ironically, though Maxwell's equations predicted electromagnetic waves, their existence was not accepted as real until nearly 25 years later, when Heinrich Hertz (1857–1894) experimentally detected electromagnetic waves.[185]

[184] Cook, *The Theory of the Electromagnetic Field.*

[185] Hertz, *Electric Waves.*

Nevertheless, Hertz was forever in awe of the equations that led him to his discovery, and here he speaks, with almost mystical reverence, of the "wisdom" that he has found in Maxwell's equations:

> One cannot escape the feeling that these mathematical formulae have an independent existence and an intelligence of their own, that they are wiser than we are, wiser even than their discoverers, that we get more out of them than was originally put into them.[186]

In the 1800s, frequencies were measured in units of "Cycles Per Second," a convention which was adopted for many decades into the twentieth century. With the development of quantum mechanics, it was realized that units of frequency often needed to be dimensionless for certain calculations to make sense. In 1930, a dimensionless frequency unit named, appropriately, "Hertz," was adopted by the International Electrotechnical Commission, officially replacing the misleading concept implied in units of "cycles per second."[187] Karl Pribram, commenting on the confusion, here suggests using the term "spectral density" in lieu of the term "frequencies":

> When we deal with a spectrum of frequencies per se, we often substitute the term "spectral density" to avoid a common confusion that Gabor pointed out in his

---

[186] E. T. Bell, *Men of Mathematics*, 16.

[187] Ruppert, *History of the International Electrotechnical Commission.*

> paper: non mathematicians, even physicists, often think of frequencies as occurring only in time, but when mathematicians deal with vibrations as in the Fourier transform, frequencies are described as composing spectra and devoid of space and time.[188]

Perhaps Pribram is reacting here to the confusion in terminology that has muddled some critics of Pribram's holographic paradigm, leading them to believe that frequency by definition (as units of cycles per second) can only exist in space–time. Here for example, Ken Wilber exhibits this fallacy, apparently misled by the term "cycles/second," in his dismissal of Pribram's holographic paradigm:

> The transform of "things" into "frequencies" is not a transform of space/time objects into space/time frequencies. Frequency does not mean "no space, no time"; it means cycles/second or space per time. To read the mathematics otherwise is more than a quantum leap; it is a leap of faith.[189]

Sine Wave and Frequency Motion of an Exoplanet

It is useful to visualize the path of a sine wave in nature. Assume we are measuring the brightness of an exoplanet circling around a distant star, and that we are viewing it almost edge-on to its planar orbit. When the planet is closest to us, it will be brightest, and when

---

[188] Pribram, *The Form Within*, 105.

[189] Wilber, *The Holographic Paradigm*, 181.

it is on the opposite side of its central star, it will appear maximally dim, being at its furthest distance from our viewing point. If we were to plot the intensity of this visual image of an exoplanet over the period during which it makes one orbital cycle, it would appear, graphed on two-dimensional Cartesian coordinates, as the sine wave in Fig. 25. Where the graph crosses the zero axis corresponds to the exoplanet being at right angles to its star and our vantage point.

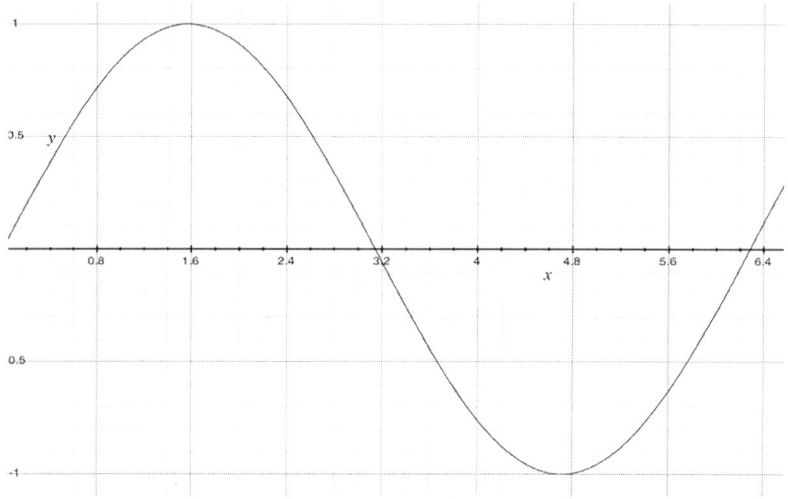

Figure *17*. A sine wave. Graphic by author.

According to the Bohr model of atomic theory, electrons spinning around their nuclei emit similar sine wave patterns as they orbit, and accordingly, space–time is replete with sinusoidal electromagnetic waves of an almost infinite number of different frequencies, corresponding to orbiting photons.[190] The frequency of radiation emitted is the inverse of the

[190] Baggott, *The Quantum Story*, 30.

orbital time; for example, an orbit that takes one second to complete would emit electromagnetic radiation of 1 Hz, while an orbit of an electron with smaller radii would exhibit a higher frequencies of radiant of energy. The closer to the center, they faster the orbit and the less time expired.

The wavelength of a sine wave is a measurement of the distance, in space, between any two points with the same phase of a frequency, such as between troughs, crests, or zero crossing in the same direction. Figure 26 illustrates the concept of wavelength, depicting the graph of a sine of wavelength λ.

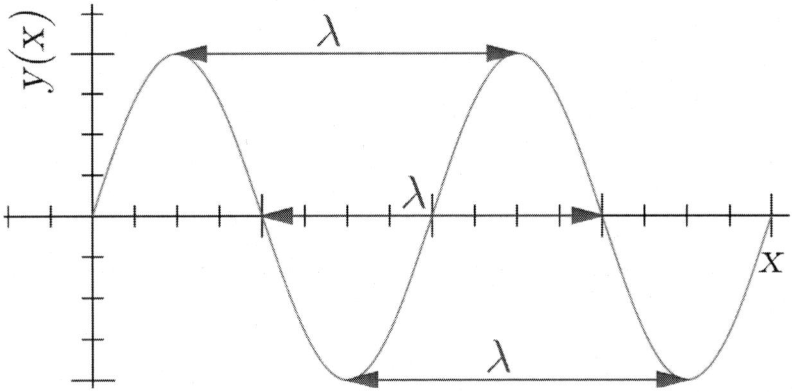

Figure *18*. Wavelength. Graphic by Richard F. Lyon (2009). Reprinted under the terms of a Creative Commons Attribution ShareAlike 3.0 Unported license.

Frequency charts of radiant electromagnetic energy are commonly depicted as a spectrum displayed on two-dimensional Cartesian coordinates with named ranges, some of which are listed in Table 3, "Ranges of the Electromagnetic Field."

Table 3

Ranges of the Electromagnetic Field

Named range	Approximate wavelengths	Frequency (Hz)
Radio waves	10^3 or 1 kilometer	10^6
Microwaves	10^0 or 1 meter	10^9
Infrared	10^{-8} meters	10^{13}
Visible light	10^{-9} meters	10^{14}
UV rays	10^{-10} meters	10^{16}
X rays	10^{-11} meters	10^{18}
Gamma rays	10^{-12} meters	10^{19}

Note. Data adapted from Benenson et al., *Handbook of Physics*, 227.

The ubiquitous word "heat" tends to confuse an actual phenomenon with an epiphenomenon. The actual the sensation termed "heat" is a measure of human neurosensory reaction to the physically radiant energy phenomenon of electromagnetic energy interacting with neurons in the skin cells. The actual energy in question is invisible electromagnetic radiation in the infrared band of the spectrum. Similarly, when we use the word "light," we tend to fall into the same potentially specious obfuscation. The term "light," like heat, pertains to the bio-sensory phenomenon of visual perception, while the primary phenomenon is actually pure radiant electromagnetic energy, precisely the same kind of energy as "heat," though of a higher frequency (a shorter wavelength) in the significantly broader electromagnetic spectrum.[191]

What is less commonly considered is that the living human body itself broadcasts a unique, narrow spectrum of electromagnetic radiation in the infrared band of radiation.[192] Wien's Law states that the maximum electromagnetic wavelength generated by a theoretically perfect "black body" at a specific given temperature is given by the following equation:

$$\lambda_{\max} = \frac{b}{T}$$

Equation 3. Wien's Law.[193] Graphic by author.

Here "b" in the equation is a constant discovered by Wilhelm Wien in 1893 and is equal to 2.897768551 x 10^3 m K, and ***T*** is the temperature.[194] If we substitute the core human body temperature of 98.6 F for ***T***, we obtain the maximum wavelength to be **9.34 micrometers**. Electromagnetic energy vibrating at this wavelength lies within the range of the infrared region, just below the range of humanly visible light frequencies. It is of interest to note that human bodies generate electromagnetic radiation at frequencies with a wavelength approximately five times smaller than the

---

[191] Feynman, Leighton, and Sands, *The Feynman Lectures on Physics*, 72.

[192] Becker and Selden, *The Body Electric: Electromagnetism and the Foundations of Life.*

[193] Feynman, Leighton, and Sands. *The Feynman Lectures on Physics, Vol. 1*, 74.

[194] Crease, *The Great Equations*, 128.

diameter of a human hair, but the approximate size of a human red blood cell.[195]

Energy vibrating in the infrared band of radiation also has a unique relationship with water. It is within the infrared range that water molecules exhibit their maximum transmittance or transparency to electromagnetic radiation, as can be seen in Fig. 27, "Water Response to Infrared Electromagnetic Radiation." A sharp dip in the absorption coefficient of atmospheric water vapor can be seen in the graph at about 10 microns on the horizontal axis. Such a dip is what communication engineers call a "passband." Earth's atmosphere exhibits a passband to radiation in the 10 micron range, the infrared. What is important to note here is that this atmospheric passband closely matches the human peak wavelength of electromagnetic radiation, previously calculated as 9.34 microns.[196]

---

[195] Becker, *Cross Currents: The Perils of Electropollution.*

[196] Warren and Brandt, "Optical Constants of Ice from the Ultraviolet to the Microwave," 1049.

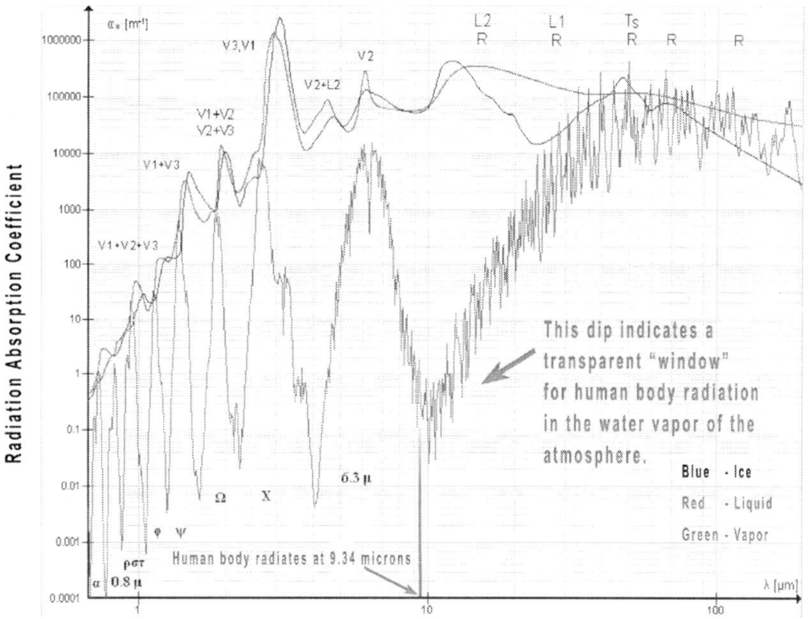

Figure *19*. Water response to infrared radiation.[197] Annotation added by author to image by Darekk2 (2012). Reprinted under the terms of Creative Commons Attribution ShareAlike 3.0 Unported. Image retrieved from Wikimedia Commons.

If the dynamics of human consciousness has evolved to broadcast and receive in this frequency range, then it would be a very convenient thing indeed that the atmosphere of our planet is transparent in this range. At 9.34 microns, the atmosphere with its gasses and water molecules does not absorb our radiant energy but allows it to pass relatively unrestricted and unattenuated in the biosphere. The passbands of the Earth's atmosphere can be seen in a graph of

[197] Water absorption spectrum; absorption coefficient for water—liquid (red line), vapor (green) and ice (blue) between 667 nm and 200 µm (red and most important part of infrared).

atmospheric electromagnetic opacity by wavelength, plotted by NASA (Fig. 28).

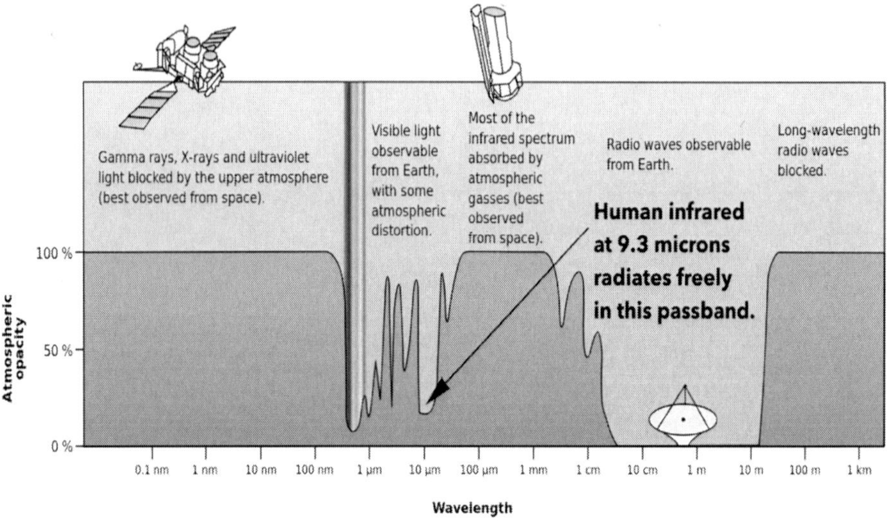

Figure *20*. Atmospheric electromagnetic opacity. Annotation added by author to NASA image by Mysid (2008). Public domain image retrieved from Wikimedia Commons.

As early as 1939 military engineers began exploiting the fact that the electromagnetic radiation emitted by human bodies is transparent to the atmosphere. This realization enabled them to develop electro-optic night vision devices specifically to observe human infrared radiation travelling through the atmosphere at wavelengths between 9 and 10 microns.[198]

Water molecules themselves, as vapor in the atmosphere, may act as magnetically resonant dipole "highways," providing infrared channels and currents in the air, even among the clouds, for electromagnetic

[198] Clark, *Photography by Infrared—Its Principles and Applications*, 78.

bio-networking around Earth's globe. Such bio-communication possibilities for human infrared radiation have been widely ignored in the neuroscience community, where such energy is understandably dismissed as "heat," or "noise," or "random thermodynamic activity," while the focus continues to be primarily upon the search for bands of human consciousness within the relatively slow, "brain-wave" frequency passband regions that are recorded by scalp-affixed electroencephalographic (EEG) sensors, generally detected in a range from 8 Hz to 50 Hz.[199]

In contrast to this low frequency range (8–50 Hz) detected by sensors on the exterior of the human scalp, the human body continuously radiates, internally and externally, in the "thermal infrared" or "long-wavelength infrared" band, which is tuned to wavelengths from 8–15 microns, in a frequency range from 27–30 THz (1 THz = 10^{12} Hz). Table 4 compares the detected ranges of human EEG frequency and human infrared frequency. The much higher frequency band of infrared as compared to EEG would translate to vast differences in information processing and storage capacity, discussed later in this chapter.

Table 4
Frequency Range: Brain Waves Versus Plasma Waves

Wave type	Frequency (Hz)
EEG "Brain Wave"	$0.8–5 \times 10^{1}$ Hz
Infrared "Plasma Wave"	$2.7–3 \times 10^{13}$ Hz

[199] Nunez, *Brain, Mind, and the Structure of Reality*, 104–5.

Note. Data from Nunez. *Brain, Mind*, 216. Author's table.

Potentially significant biological correlations can now be established. The human emission of radiation wavelengths in the 9.34 micron range not only lies within the infrared band, but also is dimensionally the same size as the prokaryote (basically, a bacteria), believed to be the first living organism on planet Earth.[200] This human electromagnetic wavelength of 9.34 microns also matches the average diameter of human blood capillaries, the smallest of the body's circulatory structures, on the average 8–10 microns in diameter; it has been estimated that there are approximately 25,000 miles of capillaries within each adult human body.[201] To appreciate the significance of the correlation of the human infrared radiation wavelength and the inner diameter of the human capillary system, it is important to understand the concept of waveguides in communication engineering.

Waveguides have been used for over a century, both commercially and in research, to channel and guide vibrating energy of specific limited frequency ranges. It was discovered late in the nineteenth century that circular metallic tubes, or hollow metal ducts, similar to A/C ventilation ducts, but much smaller, could be used to channel and guide either sound vibrations in air, or electromagnetic energy in air or vacuum. Without the waveguide, the vibrational energy field is transmitted in all directions, which may be

[200] Zimmer, "Origins: On the Origin of Eukaryotes," 667.

[201] Romanes, *Cunningham's Textbook of Anatomy*, 840.

visualized as magnetic lines, arrows emerging from a point at the center of an expanding sphere. This energy disperses outwardly, the magnetic vectored arrowheads pushing into the inside of an infinitely expanding sphere. A waveguide, however, constrains the magnetic component of the wave-front of vibrating energy to one specific linear direction, in parallel with the center of the waveguide, and thus, conceptually, the wave itself loses very little power while confined to propagate down the center of the waveguide, like a stream of water emerging from the pinprick of a large, taut, water balloon.[202]

In the mid-twentieth century, scientists began to develop microwave devices for radar and high information, bi-directional communication. In general, the inner dimension (diameter) of the waveguide must be of the same order of magnitude as the wavelength being "guided." Electrical engineers during World War II developed radar technology by using metal waveguides to channel, modulate, and detect frequencies in the microwave band.[203]

Contemporary data networks are increasingly channeled through optical waveguides using the properties of thin transparent fiber optic materials. High-speed network optical fibers currently used in the global Internet confine light to their core, most commonly using a glass fiber with a core diameter of 8–10 microns, designed for laser communication in the near infrared, the frequency band in which the ubiquitous carbon dioxide laser (CO_2 laser) emits

[202] Dorf, *The Electrical Engineering Handbook.*

[203] Ibid.

coherent radiation. The CO_2 laser is one of the most widely used lasers due to its ability to reliably operate at the highest-power efficiency currently found among laser systems.[204]

Perhaps it should not come as a surprise then that living human blood cells operate within their energy environment in the same infrared band as does our Internet data, and that the blood system is saturated with carbon dioxide gas. If conscious information processing in the human bio-system uses electromagnetic energy as a medium, then waveguide networks might also be identified within human physiology. Dimensional analysis, developed in the following section, would indicate the possibility of two systems of enclosed, tubular structures that might act as waveguides for bands of networked human electromagnetics.

A recent proposal by the University of California Los Angeles neuroscientist N. J. Woolf goes so far as to identify the site of memory storage in the locus of microtubule channels:

> Most contemporary models of consciousness posit synaptic activity as the primary basis of consciousness . . . . The view I propose is that the site of memory storage is not within the synapse, but in the sub-synaptic zone, in the microtubules of the dendrite shaft . . . the logic to this proposal is that the synapse is a channel transmitting information, much like a radio

---

[204] Kachris, Christoforos, Bergman, and Tomkos, *Optical Interconnects for Future Data Center Networks.*

receiver . . . . In this analogy, the receiving channel is the synapse or spine.[205]

Such waveguides might be a function of the blood capillary system and/or the microtubule system, as indicated in Fig. 29.

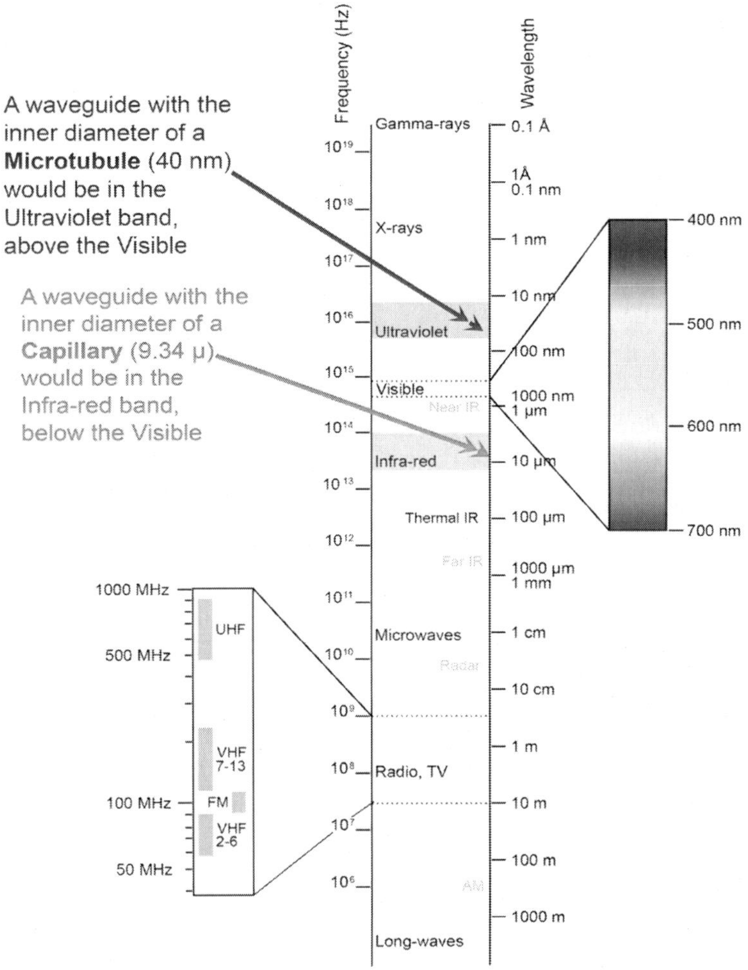

[205] Woolf, "Microtubules in the Cerebral Cortex," 83–84.

Figure *21*. Microtubules and capillaries as waveguides. Annotations by author; graphic by Jahoe (2012). Reprinted under the terms of a Creative Commons Attribution ShareAlike 2.5 Generic license. Image retrieved from Wikimedia Commons.

Information, Communication, and Meaning

Before discussing the mathematics of electromagnetic communication at the heart of holoflux theory, it is useful to examine the distinction between information, communication, and meaning. David Bohm stressed that meaning is found embedded in structure:

> Meanings are seen to be capable of being organized into ever more subtle and comprehensive over-all structures.[206]

Bohm also posits that there is a direct connection between *energy* and *meaning*, that in a sense they are aspects of the same thing:

> I would like to suggest then that the activity, virtual or actual, in the energy and in the soma is the meaning of the information, rather than to say that the information affects an entity called the mind, which in turn operates somehow on the matter of the body.[207]

As discussed in Chapter 3, Bohm went on to develop a theory he termed *active information*.[208] A

---

[206] Bohm, *Unfolding Meaning*, 75.

[207] Bohm, "Meaning and Information," 46.

similar description of information and meaning has been expressed by Rollin McCratey:

> The most basic definition of *information* is data which *in-form*, or give shape to, action or behavior, such as a message that conveys "meaning" to the recipient of a signal (Bradley & Pribram, 1998). In human language, abstract symbols like words, numbers, graphical figures, and even gestures and vocal intonations are used to encode the meaning conveyed in a message. In physiological systems, changes in chemical concentrations, the amount of biological activity, or the pattern of rhythmic activity are common means by which information is encoded in the movement of energy to inform system behavior.[209]

Communication involves three things: information, meaning, and energy transfer. For communication to occur, three things must happen: (1) an initial process in which there is an energetic transformation of meaning (understanding) into information encoded in some media, (2) an energetic transmission of the information to a receiver or receiving entity remote in space and/or time, and finally, (3) an extraction of the received information and an energetic transformation of the information (i.e., its interpretation) into meaning.

---

[208] Bohm, "Meaning and Information," 1.

[209] McCraty et al., *The Coherent Heart*, 16.

For two-way communication, the reverse must happen, the previous receiving entity encodes meaning (perhaps in response to the original meaning) into information, transmits the information through some medium (possibly the same medium, but not necessarily) to the original sender, and the information received is then translated or interpreted (understood) by the original sender. This process constitutes a full circle, a cybernetic loop, between sender and receiver.

In the spatial dimension, communication can be carried by yelling across a room. Visually, smoke signals, semaphores, or moving lips have all been used as effective means of communication. In the time dimensions, simple sticky note messages can be sent to one's self in the future (though this is one-way communication).

It was Claude Shannon (a cousin of Thomas Edison) who applied thermodynamic equations to relate the flow of thermal energy to the flow of information encoded in any medium.[210] His work led to enhanced methods of modulating information signals in different media, from electromagnetic "radio wave" energy to visible and infrared "light" energy.

Two domains, the time domain (t_d) and the frequency domain (f_d), are of essential importance in the field of information theory and signal communication. On the first page of an undergraduate electrical engineering textbook on network analysis and synthesis, Kuo states:

> In describing signals, we use the two universal languages of electrical

[210] Shannon, "A Mathematical Theory of Communication."

engineering—*time* and *frequency*. Strictly speaking, a signal is a function of time. However, the signal can be described equally well in terms of *spectral* or *frequency* information. As between any two languages, such as French and German, translation is needed to render information given in one language comprehensible in the other. Between time and frequency, the translation is effected by the *Fourier series* and the *Fourier integral*.[211]

In his book *Brain and Perception*, Pribram used almost the same words as Kuo in discussing methods by which a Fourier transform operates in mathematical physics. Here Pribram reaches the same conclusion as Kuo:

> It is reasonable to ask: What advantage does the organism gain by processing in the spectral transform domain? The answer is efficiency: The fact that correlations are *so easily achieved* by first convolving signals in the spectral domain and then, inverse transforming them into the space–time domain. Thus Fast Fourier Transform (FFT) procedures have become the basis of computerized tomography, the CT scans used by hospitals.[212]

This emphasis on the significant role of Fourier transform mathematics in electrical engineering and neurophysiological research can also be found in

---

[211] Kuo, *Network Analysis and Synthesis*, 1.

[212] Pribram, *Brain and Perception*, 73.

quantum physics (e.g., on page 1 of Bohm's own 646-page textbook, *Quantum Theory*):

> The development of the special mathematical techniques that are necessary for obtaining quantitative results in complex problems should take place, for the most part, either in a mathematics course or in a special course concerned with the mathematics of quantum theory. It seems impossible, however, to develop quantum concepts extensively without Fourier analysis. It is, therefore, presupposed that the reader is moderately familiar with Fourier analysis.[213]

Accordingly it is worthwhile to examine in detail the nature of the Fourier series and Fourier integrals, and in particular to explore their relation to imaginary numbers and the Maxwell equations that map the flow of electromagnetic energy in space–time.

Leonhard Euler and Imaginary Numbers

The two domains related by the Fourier transform, the time domain and frequency domain, had been "discovered," explored and developed as early as 1735 by Leonhard Euler (1707–1783), a Swiss mathematical genius who not only defined the concept of a "function," (Euler was the first to write $f(x)$ to denote a function driven by a variable), but also named and defined numerous mathematical symbols widely used today including the constant "pi," in Greek, the letter π; he had determined through calculation that if

[213] Bohm, *Quantum Theory*, 1.

the diameter of a circle is 1, then its circumference must be the irrational number, 3.14159265 which "for the sake of brevity" he labeled π.[214] Euler's most important work published work is a two-volume textbook written in 1748, *Introductio ad Analysin Infinitorum* (*Introduction to Infinite Analysis*).[215]

> In the *Introductio*, Euler announced the dramatic discovery of a deep connection between exponential functions, trigonometric functions, and imaginary numbers . . . Euler unveiled numerous discoveries about functions involving infinite theories . . . and proposed definitions and symbols that have since become standard, including π and e.[216]

An example can be seen here in his choice of a base for an exponential function, which he called e, created by adding up a simple infinite series:

$$e = 1 + \frac{1}{1!} + \frac{1}{2!} + \frac{1}{3!} + \frac{1}{4!} + \frac{1}{5!} + \frac{1}{6!} + \frac{1}{7!} + \frac{1}{8!} + \frac{1}{9!} + \frac{1}{10!} + \frac{1}{11!} + \frac{1}{12!} + \dots$$

This number e became Euler's base for what he called "natural logarithms," and one of the most important mathematical constants.[217] However Euler's

---

214 Crease, *The Great Equations*, 100.

215 Ibid., 98.

216 Ibid., 98–99.

greatest contribution to mathematical physics was his proof of a direct mathematical connection between the *time domain* and the *frequency domain*. The Nobel Prize winning physicist, Richard Feynman, has called Euler's discovery, "the gold standard for mathematical beauty," and "the most remarkable formula in mathematics."[218]

Mathematicians in the early eighteenth century, following Newton, sought for expressions and patterns in pure mathematics that could be shown to mirror physical phenomena observed in nature. With his photographic memory and "rare ability for concentration," Euler was able to perceive patterns in mathematical series and relationships that resulted in a prodigious number of proofs (at his death Euler had published more than 40 folio volumes on mathematics). Euler's biographer put forth his own theory on the spectacular success of this prodigy:

> The phenomenon of Euler is essentially tied to three factors: first to the gift of a possibly unique memory . . . at an advanced age, he demonstrated that he could repeat the Aeneid of Virgil from beginning to end without hesitation, and for every page in the edition he could indicate which line was the first and which the last. Secondly, his enormous mnemonic power was paired with a rare ability of concentration. Noise and hustle in his immediate vicinity barely disturbed him in his mental work: "A child

---

[217] Ibid., 100.

[218] Feynman, Leighton, and Sands, *The Feynman Lectures*, 211.

on his knees, a cat on his back, this is the way he wrote his immortal works" reports his colleague, Thiebault. The third factor of the "mystery Euler" is simply his steady, quiet work.[219]

Prior to Euler's discovery, there had never been a way to mathematically connect the timespace (t_d) dimension with the nontemporal frequency (f_d) dimension. His discovery resulted from his investigation into an obscure field of mathematics (discovered by Greek mathematicians, but first called "imaginary numbers" by Descartes by way of ridicule) in which Euler defined the square root of minus one as the letter "*i*" (for "imaginary").[220] Thus the letter *i* designates the square root of minus one:

$$i = \sqrt{-1},$$

or conversely, $\boldsymbol{i^2 = -1}$.

Euler found this imaginary number to be a great mathematical tool because multiplying "*i*" times "*i*" (i.e., squaring "*i*"), simply reverses the sign of an expression. This feature can be used to reverse the signs of various elements in a series expansion, and thus allowed Euler to develop and explore a wide range of equalities in mathematical expansion series. Use of this imaginary number as an operator, combined with

---

219 Fellmann, *Leonhard Euler*, xv.

220 While the "*i*" symbol continues to be used by mathematicians and physicists, electrical engineers use the letter "*j*" to designate imaginary numbers, primarily because they use the letter "*i*" to designate electrical current flow; Kuo, *Network Analysis and Synthesis*, 16.

his copious memory led directly to his discovery, in 1735, of what is now called Euler's Law: a direct relationship between real numbers and imaginary numbers. Depicted in Cartesian coordinates, the horizontal axis of real numbers, commonly labeled the "time" axis, is widely used in physics and mathematical charting, while the vertical, "imaginary" axis, is widely used in electrical engineering because it governs so well the mathematics of complex-plane calculations which are ubiquitous in contemporary electronic devices, during the translation of digital to analog signals and vice versa.[221]

In his *Introductio ad analysin infinitorum*, Euler announced the discovery of a connection between exponential functions, trigonometric functions, and imaginary numbers.[222] Perhaps the best way to appreciate the beauty in the discovery of Euler's Law is to go through its derivation.[223]

Derivation of Euler's Law

Euler was fascinated by what are called "infinite series," expressions by which trigonometric values can be expressed, and more importantly, calculated. By summing a series of arithmetical terms, the value of a sin or cos of an angle can be calculated to any degree of precision by approximating them as the sum of a series

---

221 Kuo, *Network Analysis and Synthesis*, 77.

222 Crease, *The Great Equations*, 99.

223 Material in the following section is taken from Crease, "Euler's Equation," in *The Great Equations*, 91–110; mathematical expressions in this section are by author.

of arithmetic values. Obviously an entire infinite series could not be calculated, but approximations were acceptable because it allowed the creation of large tables of values that were in widespread demand for use by engineers in the sixteenth century throughout Europe.[224]

Euler spent countless hours developing various infinite series, in the attempt to discover connections and relationships among the patterns. Two of these series especially intrigued him, the Scottish mathematician Colin Maclaurin's infinite series expansion of the sine and the cosine functions, both shown partially expanded as follows:

$$\sin x = x - \frac{x^3}{3!} + \frac{x^5}{5!} - \frac{x^7}{7!} + \frac{x^9}{9!} - \frac{x^{11}}{11!}\ldots$$

$$\cos x = 1 - \frac{x^2}{2!} + \frac{x^4}{4!} - \frac{x^6}{6!} + \frac{x^8}{8!} - \frac{x^{10}}{10!}\ldots$$

Note that the "factorial sign" (!) represents the product of all numbers starting from the indicated number down to 1, (i.e., 3! = 3 x 2 x 1 = 6), and also that the ellipsis (...) in each of the expressions above indicate an infinite series of additional factors, which follow the same pattern of progression. When Euler discovered an infinite series expansion for his own natural logarithm (the base used widely in nuclear physics, was later named "e" for Euler), he noticed how the Maclaurin's series seemed similar to the pattern of expansion of his own discovery of the e^X expansion:

[224] Crease, *The Great Equation*, 100.

$$e^x = 1 + x + \frac{x^2}{2!} + \frac{x^3}{3!} + \frac{x^4}{4!} + \frac{x^5}{5!} + \frac{x^6}{6!} \ldots$$

Euler noticed that this was strangely similar to the Maclaurin expansions for cos x and sin x added together; the result of such an addition (shown below, the sum of the two series at the top of this page) is identical to the expansion of e^x, except where minus signs appear.

$$\sin x + \cos x = 1 + x - \frac{x^2}{2!} - \frac{x^3}{3!} + \frac{x^4}{4!} + \frac{x^5}{5!} - \frac{x^6}{6!} - \frac{x^7}{7!} + \frac{x^8}{8!} + \ldots$$

For many months Euler struggled to unlock the secret of what seemed to be a remarkably close connection. Finally, it dawned on him that the relationship might find closure if he could find a way to use his imaginary number "*i*," the square root of minus 1 to change signs.

Substituting "*ix*" for "*x*" everywhere in his equation (an allowable substitution, since by the rules of algebra any valid expression can be substituted for "*x*"), he found that wherever "i^2" could be identified and factored out, the sign for that expression would reverse, and he quickly saw that his expansion would then more closely match that of the expansion of *cos x* + *sin x*.

To recapitulate, here again is Euler's original infinite expansion of the natural logarithm that he had previously discovered:

$$e^x = 1 + x + \frac{x^2}{2!} + \frac{x^3}{3!} + \frac{x^4}{4!} + \frac{x^5}{5!} + \frac{x^6}{6!}...$$

Wherever there is an "*x*" in the above equation, Euler substituted "*ix*,"

$$e^{ix} = 1 + ix + \frac{(ix)^2}{2!} + \frac{(ix)^3}{3!}$$
$$+ \frac{(ix)^4}{4!} + \frac{(ix)^5}{5!} + \frac{(ix)^6}{6!}...$$

Wherever he found an "i^2" he converted it to its computed value of "-1":

$$e^{ix} = 1 + ix - \frac{x^2}{2!} - \frac{ix^3}{3!} + \frac{x^4}{4!} + \frac{ix^5}{5!} + ...$$

Rearranging the results slightly, we can see the original patterns of $\cos x$ and $\sin x$ expansion on the left, now seen as equal to Euler's e^{ix} (on the right): this equation is known as "Euler's Law":

$$e^{ix} = \left(1 - \frac{x^2}{2!} + \frac{x^4}{4!} + ...\right) + (ix - \frac{ix^3}{3!} + \frac{ix^5}{5!} + ...)$$

Now the i can be factored out from the right side of the equation:

$$e^{ix} = \left(1 - \frac{x^2}{2!} + \frac{x^4}{4!} + ...\right) + i * (x - \frac{x^3}{3!} + \frac{x^5}{5!} + ...)$$

The expressions now within the brackets are simply the

Maclaurin expansion series for $\cos x$ and $\sin x$, and the entire expression can now be written as:

$$e^{ix} = \cos x + i * \sin x$$

The amazement generated by this discovery was that for the first time a solid, derivable mathematical link had been established between e,the natural exponential function, the imaginary number i, and trigonometric geometry; the discovery was published in 1735 by Euler (Eqn. 4).

$$e^{ix} = \cos x + i * \sin x$$

Equation 4. Euler's Law.[225] *Graphic by author.*

Notice how both sides of Euler's Law contain both real numbers (***x***) and imaginary numbers (***i***). Euler's Law maps the intersection of the axis of real numbers with the angular frequency axis of imaginary numbers on a single intersecting plane, allowing us to model mathematically the real world of timespace/frequency phenomena of quantum electrodynamics. Euler's Law has since become the basis of electronic communication technology within which our society is currently enmeshed.[226]

During the eighteenth century this field, "complex mathematics," remained an obscurity, until "rediscovered" by physicists such as Maxwell, Tesla, and Marconi in the late nineteenth century; the

---

225 Crease, *The Great Equations*, 102.

226 Blakeslee, *The Radio Amateur's Handbook*, 23.

mathematics of imaginary numbers became an essential tool for modeling an invisible electromagnetic reality, and more importantly, provided a direct way of calculating, predicting, and modulating the electromagnetic energy oscillations and waves associated with the newly emerging technologies of alternating current. [227] Nevertheless, it was Euler who first established the mathematical beachhead into the Real–Imaginary domain, and this relationship was eventually found, perhaps mysteriously, to mirror electromagnetic realities in space–time.[228]

The Integral Operator and the Fourier Transform

Seventy-four years after Euler published his famous theorem, Jean Baptist Joseph Fourier (1768–1830), expanding upon Euler's mathematical discovery, published what came to be known as the Fourier series and the Fourier Transform, in a paper entitled *The Analytic Theory of Heat.*[229] Before describing the Fourier Transform, it is useful to begin by examining the word "integral" through the perspective of history. The word integral itself stems from a mathematical sign, a convention that was first introduced by Gottfried Wilhelm von Leibniz (1646–1716) as a stylized, elongated letter *S* which he used as shorthand for the Latin word *Summa* (translates "sum"

---

[227] Kuo, *Network Analysis and Synthesis*, 14.

[228] Crease, *The Great Equation*, 108.

[229] Fourier, *The Analytic Theory of Heat.*

or "total") to denote something that originates from a summation.[230]

Both Leibniz and Isaac Newton (1643–1727) seem to have developed the fundamental concepts of integral calculus simultaneously, but Newton's notation of using a vertical mark to indicate summation became the cause of considerable notational confusion (a vertical slash is "easily confused with the numeral, '1,' or with a bracket, or with the letter 'I'"), and over the next 15 years, Leibniz's elongated letter *S* found preference among mathematicians to denote "summation."[231] The word "integral" itself was coined by Jacob Bernoulli (1654–1705), so that the *S* notation could more easily be discussed in mathematical conversation and in lectures.[232]

However it was not until 1822, when Jean Baptiste Joseph Fourier introduced the notation of upper and lower limits to the integral operator symbol, that the full power of integral calculus was unleashed, making possible the development of his "Fourier series."[233]

A full century after Euler, it was Fourier, working at the time as Napoleon's governor of Egypt, who managed to link the real and imaginary axes by

---

230 Crease, *The Great Equations*, 96.

231 Browder, *Mathematical Analysis: An Introduction*, 121.

232 Livio, *The Golden Ratio: The Story of Phi*.

233 Feynman, Leighton, and Sands, *The Feynman Lectures on Physics*, 271.

building upon Euler's Theorem. While Euler had provided the initial link between the frequency and timespace domains, it was Fourier, during his experimental investigation of heat flow, who developed the mathematics to model the thermodynamic properties of energy, and was able to derive a mathematical operation of integral calculus that expressed accurately the energy transformations between a time domain (t_d) and a frequency domain (f_d).[234]

However, the mathematics expressed in the Fourier transform can also be understood as a link between a physics of space–time (t_d) and an actual, ontologically real frequency domain (f_d).

The Fourier transforms are shown in Equation 5:

$$f(t) = \int_{-\infty}^{+\infty} X(F) e^{j2\pi}$$

Fourier integral transform of a continuous time function into the frequency domain (f_d).

$$f(F) = \int_{-\infty}^{+\infty} x(t) e^{-j2}$$

Fourier integral transform of a continuous frequency function into the time domain (t_d).

Equation 5. Fourier Transform/Inverse Transform.[235]

These Fourier transform expressions indicate that any arbitrary function in the timespace domain, $f(t)$, can be transformed into and expressed by an infinite series of frequency spectra functions $X(F)$ in the

[234] Ibid., 286.

[235] Stein and Shakarchi, *Fourier Analysis*, 134–36.

frequency domain, and conversely, that any arbitrary function in the frequency domain, $f(F)$ can be transformed into and expressed by an infinite series of time functions, $x(t)$.

These Fourier transforms are themselves derived from an underlying series of alternate pure sine and pure cosine waves, as depicted in Equation 6.

$$f(t) = a_0 + \sum_{n=1}^{\infty}\left(a_n \cos\frac{n\pi t}{L} + b_n \sin\frac{n\pi t}{L}\right)$$

Equation 6. The Fourier Series.[236]

A century after Fourier's death, Norbert Wiener made use of Fourier's transform to model and analyze brain waves, and he was able to detect frequencies, centered within different spatial locations on the cortex, that exhibited auto-correlation. Specific frequencies were found to be attracting one another toward an intermediate frequency, thus exhibiting resonance or "self tuning" within a narrow range of the frequency domain (f_d).[237] This discovery led Wiener to conjecture that the infrared band of electromagnetic flux may be the loci of "self-organizing systems."[238]

> We thus see that a nonlinear interaction causing the attraction of frequency can generate a *self-organizing system*, as it does in the case of the brain waves we have

[236] Kuo, *Network Analysis and Synthesis*, 40.

[237] Wiener, *Cybernetics: Control and Communication in Animal and Machine*, 198.

[238] Ibid.

> discussed . . . . This possibility of self-organization is by no means limited to the very low frequency of these two phenomena. Consider self-organizing systems at the frequency level, say, of infrared light.[239]

Wiener goes on to discuss such possibilities in biology, where he focuses upon the problems of communication at molecular and primitive cellular levels, specifically on the problem of how substances produce cancer by reproducing themselves to mimic pre-existing normal local cells. Molecules do not simply pass notes to one another, and they do not have eyes, so how do they perceive and how do they communicate? Wiener conjectures:

> The usual explanation given is that one molecule of these substances acts as a template according to which the constituent's smaller molecules lay themselves down and unite into a similar macromolecule. However, an entirely possible way of describing such forces is that the active bearer of the specificity of a molecule may lie *in the frequency pattern of its molecular radiation*, an important part of which may lie *in infrared electromagnetic frequencies* or even lower. It is quite possible that this phenomenon may be regarded as a sort of attractive interaction of frequency.[240]

---

[239] Ibid., 202.

[240] Ibid. Italics added.

At the end of his classic book on cybernetics, in the chapter, "Brain Waves and Self-Organizing Systems," Wiener suggests further possible studies to "throw light on the validity of my hypothesis concerning brain waves."[241] Wiener goes on to describe the widespread observations of seemingly simultaneous behavior of groups of living organisms such as crickets, tree frogs, or fish, activity that might be attributable to simultaneous synchronization of neuronal networks through resonant tuning within the frequency domain (f_d):

> It has often been supposed that the fireflies in a tree flash in unison . . .
> I have heard it stated that in the case of some of the fireflies of Southeastern Asia this phenomenon is so marked that it can scarcely be put down to illusion . . . . Could not the same supposed phenomenon of the pulling together of frequencies take place? However this process occurs, it is a dynamic process and involves forces or their equivalent.[242]

The Real–Imaginary Domain and the Mandelbrot Set

In early explorations of the real–imaginary domain model in the nineteenth century, the Danish mathematician Caspar Wessel, and the mathematical physicist and astronomer Johann Carl Friedrich Gauss independently discovered that two-dimensional

---

[241] Ibid., 200.

[242] Ibid.

Cartesian plots or graphs could be made of the real–imaginary axes, with one axis of real numbers (traditionally illustrated by a straight horizontal line with values increasing from left to right) and a vertical axis of imaginary numbers drawn at a 90-degree angle to the real number axis.[243] In the twentieth century this two-dimensional model has been adopted by electrical engineers to model, analyze, and solve complex problems dealing with transformations between the time (t_d) and frequency (f_d), domains in electrical power and communication engineering.[244] The concept remains at the essential core of engineering calculations for transforming electromagnetic energy into information encoded in light, sound, and multidimensional images.[245]

Perhaps the most fascinating tool for exploring the nature of the interface between the Real–Imaginary domains was developed in 1980 when the Polish–American engineer, Benoit Mandelbrot, created the software to plot an actual image of the two-dimensional interface between space–time and frequency domains close to the origin (defined as the intersecting point where the Real axis equals zero and the Imaginary axis equals zero). His initial impression, upon seeing the first image, was that the computer program had malfunctioned.[246] Subsequent computer

---

243 E. T. Bell, *Men of Mathematics*.

244 Kuo, *Network Analysis and Synthesis*.

245 Ibid., 389.

246 Mandelbrot, “Fractals and the Rebirth of Iteration

plots assured him that these visual patterns were truly there. Images of this region about the time frequency origin have gained interest worldwide and the region itself has come to be known as the Mandelbrot set (fig. 30). The English mathematical physicist Sir Roger Penrose was so taken by the resulting images that he described them with a sense of almost reverential awe:

> The Mandelbrot set is not an invention of the human mind: it was a discovery. Just like Mount Everest, the Mandelbrot set is just *there*![247]

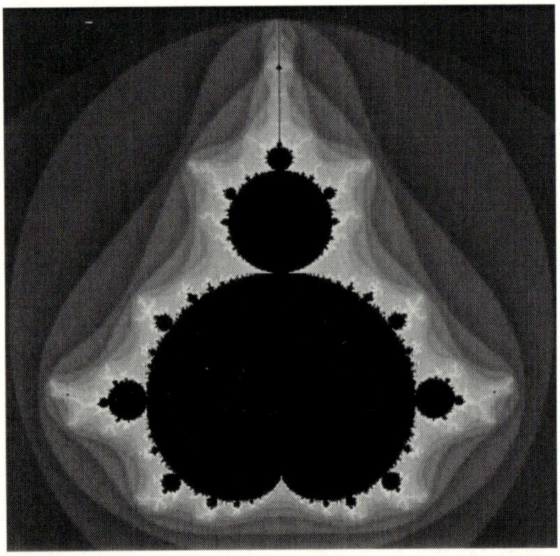

Figure 22. Mandelbrot: Origin of real–imaginary axis. Graphic by author.

The Mandelbrot set exhibits remarkable properties. As calculations are done on ever smaller regions on the time–frequency plane, the images

Theory," 151.

[247] Penrose, *Concerning Computers, Minds and the Laws of Physics*, 124.

appear similar but never completely repeatable, and the viewer begins to sense some sort of biological shapes emerging from this strange world of pure mathematical being. Penrose goes on to say, "The very system of complex numbers has a profound and timeless reality which goes quite beyond the mental constructions of any particular mathematician."[248]

Exploitation of the basic properties of this mysterious real–imaginary domain have led directly to twenty-first century digital device systems, and accordingly, it is entirely conceivable that they might also be involved in some space-time-frequency spectrum of the psyche, as theorized by Jung: "Psychic processes therefore behave like a scale along which consciousness 'slides.'"[249]

> Using the analogy of the spectrum, we could compare the lowering of unconscious contents to a displacement towards the red end of the colour band, a comparison which is especially edifying in that red, the blood colour, has always signified emotion and instinct . . . . The dynamism of instinct is lodged as it were in the infra-red part of the spectrum . . . . Psychologically, however, the archetype as a spiritual goal toward which the whole nature of man strives . . . as such is a psychoid factor that belongs, as it were, to the invisible, ultraviolet end of the psychic spectrum.[250]

---

[248] Ibid., 127.

[249] Jung, "On the Nature of the Psyche," 207.

[250] Ibid., 187–213.

Using Fourier analysis and the Fourier transform, signals can be described either in the time domain or the frequency domain, and they can be quickly transformed back and forth between the two domains. It can be said that they are two different aspects of one and the same thing, energy signals encoded with information content, but potentially existing in either one, or both, of two categorically distinct dimensions. Jung observes the relationship of psyche to matter with the same pattern, noting that they are two different modes of one and the same thing:

> Since psyche and matter are contained in one and the same world, and moreover are in continuous contact with one another and ultimately rest on irrepresentable, transcendental factors, it is not only possible, but also fairly probable, even, that psyche and matter are two different aspects of one and the same thing.[251]

If we assume that the energy of the psyche expresses itself in some category of spectral frequency range within the human biosystem, as Jung has speculated, then a mathematical model of psychic energy activity might be found in the Fourier transform. Humans in contemporary society experience the fruits of the mathematics of the Fourier transform reliably every time a digital communication device is operated.[252]

---

[251] Jung, "On the Nature of the Psyche," 215.

[252] Brigham, "The Fast Fourier Transform."

For example, signal processing chips in cell phones are encoded with what are called fast Fourier transform algorithms, which filter audio voice speech frequency patterns in the space–time domain into a relatively small set of pure frequencies, the individual amplitudes of which are then digitized and transmitted. On the receiving end, a reverse fast Fourier transform unpacks the frequency domain data, transforming them into space–time frequency waves of low voltage circuits, which then drive the vibrating speaker devices by which humans "hear" the re-created audio space–time spectral energy.[253]

Pribram describes this same process as it occurs in the brain, while emphasizing the difference between waves and spectra:

> Waves occur in space–time. Spectra do not. Spectra result from the interference among waves—and, by Fourier's insightful technique, the locus of interference can be specified as a number that indicates the "height" of the wave at that point. The number is represented by a Gabor function (or other wavelet.) Frequency in space–time has been converted to spectral density in the transform domain.[254]

Resonance in the Frequency Domain: Faraday Waves

An understanding of the phenomenon of *resonance* is of fundamental importance in any physical approach to communication and information transfer. Rollin McCraty, in discussing his research

[253] Ibid.

[254] Pribram, *The Form Within*, 109.

measuring the energy frequencies transmitted by the beating human heart, describes the phenomenon of resonance. He notes that remarkably large vibrations are observed when specific external frequencies occur close to what have been termed *resonant* frequencies of a beating heart:

> In physics, resonance refers to a phenomenon whereby an unusually large vibration is produced in a system in response to a stimulus whose frequency is identical or nearly identical to the natural vibratory frequency of the system. The frequency of the vibration produced in such a state is said to be the *resonant frequency* of the system.[255]

Michael Faraday (1791–1867) discovered resonance in what he believed to be invisible waves of electricity and magnetism vibrating in what he thought to be a solid plenum, the "ether." Faraday not only discovered the laws of electromagnetic induction, and that magnetism can affect light, but also explored the resonance phenomenon in physical systems.[256]

Faraday reported, in 1831 at the Royal Society of London, an unexpected observation during his experiments with the properties of liquids on a vibrating surface. While experimenting with thin films of liquid on the flat heads of vertically vibrating pistons, he observed distinct geometric patterns emerging on the surface of the liquid at specific fixed

---

[255] McCraty, *The Energetic Heart*, 4.

[256] Crease, *The Great Equations*, 135.

frequencies. The image below (in fig. 31, "Faraday Wave Patterns") presents twelve examples of these resonant patterns, corresponding to 12 different resonant frequencies in the vibrating fluid; the images are composed of micro-scale polystyrene beads, 200 microns in diameter, each floating on a thin layer of liquid. The distinctly unique geometric surface patterns are formed by the accretion of beads at nodal intersections, where the resonance of standing waves are at a minimum, and are similar to Prigogine's "valleys of minimum potential attractors," in his dissipative structure experiments.[257] It is interesting to observe here naturally occurring resonance patterns that appear identical to abstract symbols used in various human cultures throughout history, such as the circle, the cross, and the swastika.[258]

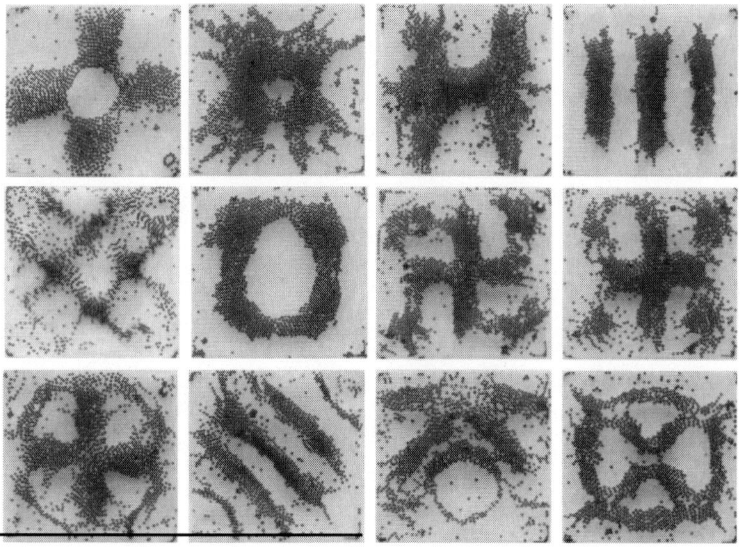

[257] Prigogine, *Thermodynamics: From Heat Engines to Dissipative Structures.*

[258] Aigner, *The Swastika Symbol in Navajo Textiles.*

Figure 23. Faraday wave patterns. Graphic by Faraday Telsa (2012). Reprinted under the terms of a Creative Commons Attribution ShareAlike 3.0 Unported license. Image retrieved from Wikimedia Commons.

Since Faraday's time, the concept of resonance has become conventionally associated with music and acoustic phenomena, as in "he had a resonant voice," implying some perceived richness in harmonic acoustic overtones. However in physics, the resonance effect is found to have a more precise meaning, as applied to mechanical and electromagnetic systems, and in fact the phenomenon of resonance has become a fundamental tool in communication theory and the design of electrical circuits.[259] Resonance is a phenomena characteristic of physical objects extended in space–time, in which objects often exhibit remarkable frequency sensitivity to particular input frequencies, flowing in, through, and around the object, which approach the object's natural frequency (Fig. 32). Perfect resonance occurs where the input frequency and the natural frequency are one.[260]

[259] Dorf, *The Electrical Engineering Handbook*, 41.

[260] Feynman, Leighton, and Sands, *The Feynman Lectures on Physics*, 78.

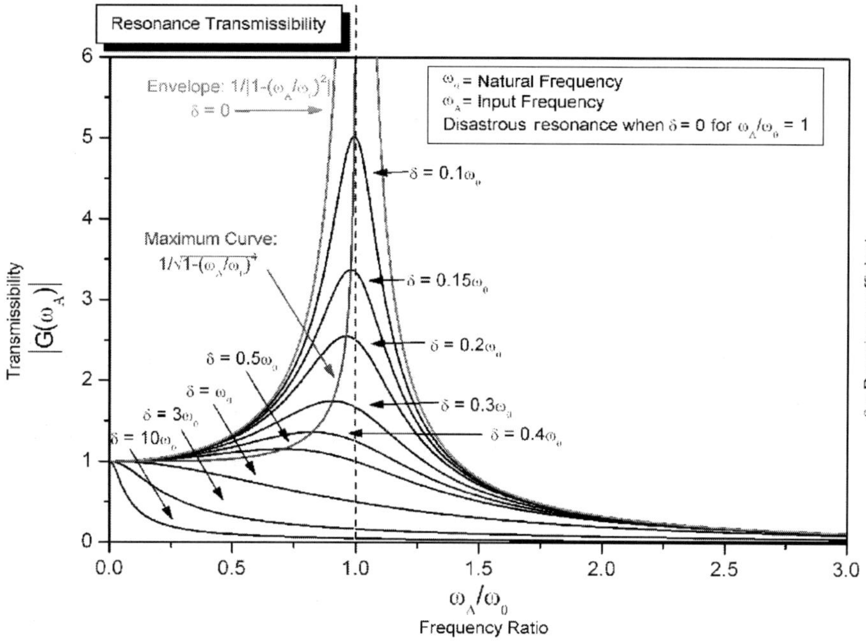

Figure *24*. Resonance effect. Graphic by Amuecki (2008). Reprinted under the terms of a Creative Commons Attribution ShareAlike 3.0 Unported license. Image retrieved from Wikimedia Commons.

This principle of resonance governs the design of antennas in electromagnetic transmitting and receiving systems. The goal of antenna design is to construct an antenna that is maximally resonant with an incoming electromagnetic signal; the input frequency and the natural frequency of the antenna must be the same frequency for resonance to occur:

> A fundamental form of antenna is a single wire whose length is approximately equal to half the transmitting wavelength. It is the unit from which many more-complex forms of antennas are constructed. It is known as a *dipole antenna*.[261]

As can be seen in the figure of a sine wave (Fig. 25), the distance between the maximum and minimum values of the sine wave, the maximum distance between peak and trough, is one half of the complete waveform cycle, or exactly the length of a dipole antenna which has been constructed to match, "tuned" to that sine wave frequency.[262]

Resonant electronic components are ubiquitous building blocks in the operation of electronic devices, and resonant mechanical systems frequently exist in the natural world. Resonant conditions in a system exhibit one of two behaviors; the resonance either (1) restricts or blocks the flow of energy into the system, or (2) enhances the flow of energy into the system. In signal analysis the resonance conditions are mathematically termed "poles and zeros"; poles are seen "as being the roots of the denominator" of a signal function, while zeros are "defined as the roots of the numerator" of a signal function.[263] A *resonant pole* in the numerator takes the signal power to infinity, while a *resonant zero* in the denominator, numerically approaching infinity, causes signal power to approach zero.[264] One example of the phenomenon of a resonant pole can be seen in the effect that has been demonstrated by an opera singer acoustically shattering a wine glass by sustaining a high frequency

---

[261] Blakeslee, *The Radio Amateur's Handbook*, 580.

[262] Ibid.

[263] Kuo, *Network Analysis and Synthesis*, 183.

[264] Ibid.

note which happens to be the natural resonant frequency of the crystal (a *pole* in the numerator of a signal function).[265]

Superposition in the Frequency Domain

Also of importance to an understanding of signal communication in the frequency domain is the concept of *superposition*. The combination, or superposition, of multiple frequencies in the frequency domain can create unique patterns in the space–time domain. The Berkeley physicist Henry Stapp explains the importance of superposition in his model of brain dynamics:

> The rudiments of brain dynamics must be understood. In a normal computing machine, the currently active information is stored in a generally small number of registers. But in the brain a huge number of separate patterns of neural excitations can be present at one time. These patterns can become correlated to stimuli and response, and can mediate the behavior of the organism . . . . The main postulate of the model is that every conscious event is the psychological counterpart of a certain special kind of Heisenberg event in the brain, namely an event that actualizes a pattern of neuronal activity that constitutes a *representation* of this general kind. However, any such representation must be *formed before it can be selected*: the representation must be constructed by

---

265 Schrock, "An Opera Singer's Piercing Voice Can Shatter Glass."

> unconscious brain activity before it can be actualized. During this preliminary mechanical phase, a ***superposition*** of many such representations must inevitably be generated. During the subsequent actualization phase, only *one* of these representations will be selected.[266]

An example of the superposition principle can be seen in Fig. 33 where the sum of a number of sine waves, shown at the bottom of the figure, results in the "sawtooth wave" seen in the top of the figure.

At the bottom half of the figure are five different frequencies mapped on Cartesian coordinates; the combination of these frequencies, superimposed and translated through a Fourier transform, reveals the sawtooth pattern seen in the upper half of the figure in the time domain.[267]

---

266 Stapp, *Mind, Matter and Quantum Mechanics*, 25. Emphasis added in bold by author.

267 Howell, *Principles of Fourier Analysis.*

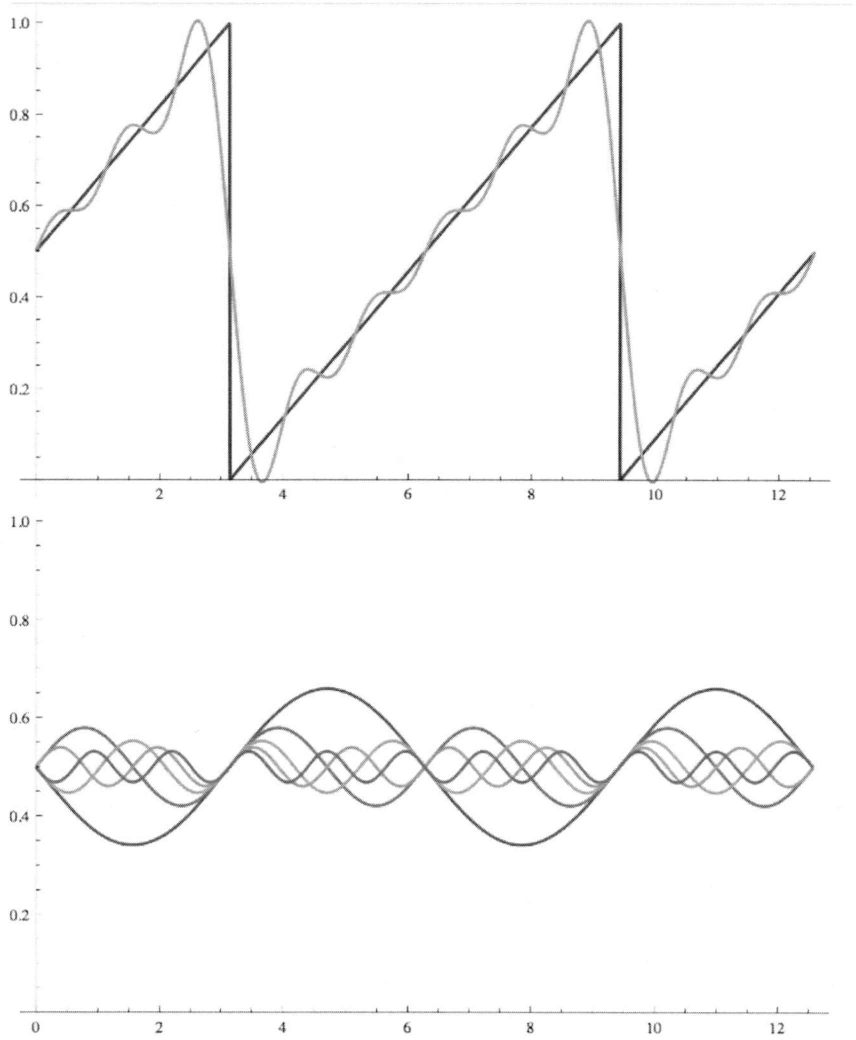

Figure *25*. Sawtooth wave pattern created from five frequencies. Graphic by Differenxe (2010). Reprinted under the terms of a Creative Commons Attribution ShareAlike 3.0 Unported license. Image retrieved from Wikimedia Commons.

At this point it is useful to explore a key question regarding the ontological reality of mathematical functions operating in the frequency domain: does the frequency domain have an ontological reality, does it exist, or is it simply a mathematical fiction, a fabrication? The many working products of electrical engineering indicate that frequency domain calculations do mirror something real in the cosmos, and experimental evidence that the Schrödinger wavefunction has its own reality gives credence to regarding the frequency domain as also having a cosmological existence beyond that of space–time.

Reality of Schrödinger Wavefunction (ψ)

The applicability of the Schrödinger wavefunction (from which Bohm's quantum potential wavefunction was subsequently derived) may be said to be at the heart of quantum mechanics, but the ontological reality of this wavefunction has not until recently been tested.[268] In 2015, a research team in Australia, at the Centre for Quantum Computer and Communication Technology at the University of Queensland, published results of a series of experiments to explore the nature of the primary mathematical tool used in quantum mechanics, Schrödinger's quantum wavefunction. In describing their research objective, they state, "It is still unclear what the quantum wavefunction actually is: does it merely represent our limited knowledge of a system, or is it an element of reality?"[269] Analyzing four-

---

[268] See Equation 1, "Equation of the Quantum Potential."

[269] Ringbauer et al., "Measurements on the Reality of the Wavefunction," 249.

dimensional states of photons, their experiment *detected the reality of the wavefunction,* and experimental results strengthened the probability that the wavefunction is part of the ontological cosmos, not simply a conceptual tool:

> Our results suggest that, if there is objective reality, the wavefunction corresponds to this reality. These results strengthen the view that the entire wavefunction should be real. The only alternative is to adopt more unorthodox concepts such as backwards-in-time causation, or to completely abandon any notion of objective reality.[270]

These results, strengthening the premise that Schrödinger's wavefunction is ontologically real, also increase the likelihood that Bohm's quantum potential wavefunction, derived directly from Schrödinger's equation, is similarly "real," and that the frequency domain of electrical engineering is also "real." The implications support Bohm's vision, described here, of information being carried by an explicate level of activity (space–time), while being seen from "a finer, subtler implicate level":

> In analogy to what has been said about human experiences, the particles constituting matter in general may be considered to represent a more gross (explicate) somatic level of activity, while the Schrödinger wave field corresponds to a finer, subtler, more implicate and "mind-like" level. In human experience however, it has been proposed that each "mind-like"

[270] Ibid.

> level can be regarded as a somatic bearer of form when seen from a yet finer and more subtle level. This would imply firstly that the information represented by the Schrödinger wave field is being "carried" by a finer and subtler level of matter that has not yet been revealed more directly. But even more important, it also implies that there may be a finer and more subtle level of information that guides the Schrödinger field, as the information on the Schrodinger field guides the particles.[271]

As discussed in Chapter 4, it is Bohm's quantum potential (Q) of "active information" that provides the guidance that subtly influences the larger effects of the Schrödinger quantum wavefunction (ψ) in the much larger world of particle physics.[272] It is the thesis of this book, backed by current quantum theory, that Bohm's quantum potential manifests as a holoflux process of frequency-phase energy within the frequency domain (the implicate order), resonating with and so influencing electromagnetic energy waves within the space–time domain (the explicate order).

Holoflux in The Holoplenum

Holoflux, as defined in this thesis, is a transformational process of modulated energy flowing into, out from, and within a holoplenum of holospheres in frequencies as high as 10^{44} Hz. This holoflux manifests a flowing field of consciousness, the

[271] Bohm, "Meaning and Information," 59.

[272] Bohm and Peat, *Science, Order, and Creativity*, 80–81.

modulation of which is information embedded in the waves and frequencies of space–time energy. Within each holosphere, below the Planck length of 10^{-35} meters—though the words "within" and "below" are clearly inappropriate for a region that is nonlocal—lies what Bohm calls the *implicate order*.

Holoflux can be regarded as a type of plasma energy that manifests consciousness in two different modes within two different domains: (1) In the space–time domain, where it manifests in electromagnetic waves, the holoflux process consists of reflective, time-sequenced snapshots of awareness, like glass beads on a thread, and (2) In the frequency domain of the nonlocal, the implicate order, holoflux manifests as immediate awareness (a timeless-sense, the "consciousness-without-an-object" experienced by the mathematician–philosopher Franklin Merrell-Wolf).[273]

Holoflux, as a total manifestation, is a dual process of reflective consciousness and transcendent awareness. Communication, and consciousness flow between and within the two domains: the implicate and the explicate. Another way of regarding holoflux is to see it as a two-way moving flux: (1) It is an explicate energy (E) moving *inward* into the implicate order at the holospheric event horizon of a quantum black hole, and (2) Moving in the opposite direction, it is an implicate energy (Q) moving *outward* as quantum potential from the event horizon, as holoflux energy transforms from out of the implicate order into the explicate order. My thesis rests upon the assumptions that Bohm's implicate order is real, that the frequency

[273] Merrell-Wolff, *The Philosophy of Consciousness Without an Object*.

domain is real, and that his active information wavefunction is real, affecting our world of "particles" as it springs from the virtually infinite nonlocal information holoflux within the implicate order.

Like Bohm, electrical engineers separate the world into two domains, which they term the *complex-frequency domain* and the *time domain.* It is our normally perceived space–time that electrical engineers identify as the *time domain*, but it is in the *frequency domain* that they see the operation of communication signals.[274]

Franklin F. Kuo, an American computer scientist and communications engineer at Bell Laboratories, wrote an electrical engineering textbook, developing the mathematics of communication in *Network Analysis and Synthesis*. In the first chapter of his classic textbook, Kuo describes the two domains, and the mathematical advantage of creating and receiving communication signals within the *frequency* domain as opposed to the *space–time* domain:

> In the *complex-frequency domain*, the voltage–current relationships for the elements are expressed in *algebraic* equations. Algebraic equations are, in most cases, more easily solved than differential equations. Herein lies the raison *d'être* for describing signals and networks in the *frequency domain* as well as in the *time domain*.[275]

---

[274] Kuo, *Network Analysis and Synthesis*, 13.

[275] Ibid.

These two orders of electrical engineering theory can be seen to reflect David Bohm's explicate and implicate orders. Bohm's explicate order is associated with the engineering *time* dimension through which signals are transmitted and received sequentially in time, while Bohm's implicate order can be identified with the electrical engineering *frequency* dimension as a timeless, spaceless domain of pure frequency-phase information.

It has been discovered that these two domains are strongly linked, mathematically, with the operation of the Fourier transform. As Kuo says, "between time and frequency, the translation is effected by the Fourier series."[276] As we have seen in Chapter 4, the Fourier series has also become the bedrock of quantum theoretical calculations, as expressed by David Bohm in the opening of his textbook, *Quantum Theory*:

> It seems impossible to develop quantum concepts extensively without Fourier analysis.[277]

Our thesis, then, is that Kuo's frequency domain, bridged to time by Fourier's mathematics in our twenty-first-century information technology, has an ontological reality, and that there does exist an actual frequency domain, which we also identify with Bohm's hypothesized implicate order.

---

[276] Ibid., 3.

[277] Bohm, *Quantum Theory*, 1.

The Dynamics of Holoflux

Having established the basic architecture of the holoflux theory, it is possible to go further by imagining the cycle nature of holoflux information as it manifests in and between the two domains, from a Bohmian perspective.

During each inward (implicate) interval (of 10^{-44} seconds), Bohm's quantum potential undergoes reconfiguration calculations within the nonlocal frequency domain. These internal calculations use information that has been feeding inwardly from the explicate (space–time) order during previous "cycles." The sub-quantum calculations within the implicate yield solutions involving frequency superposition and fractal pattern formation, likely far beyond our current mathematical capabilities, within the frequency domain. The results of this holoflux superposition-calculation are then posted into space–time, and incorporated into the outflowing modulation of holoflux as the quantum potential *Q*, of active information. These manifest in waveforms entering space–time as fields of modulated electromagnetic energy.

Conversely, information flowing within space–time, not only travels outward, but moves (in-folds and enfolds) towards the center, back into the implicate order, to be captured by the quantum black hole of the Planck holosphere at its event horizon. Such enfolding information is immediately available within the implicate order/frequency domain for superposition-calculation of the next out-going quantum wave into space–time. The resulting outgoing wave of quantum potential (*Q*) of "active information" is encoded with information to nudge every photon in the universe in space–time.

Holoplenum Plasma Display

It is possible to comprehend the emerging projection of space–time reality from the holoplenum by expanding upon the metaphor of a flat-panel plasma display. The human visual threshold for detecting separate images lies somewhere between 10 to 12 separate images per second, and accordingly the industry standard in the motion picture industry is 24 frames per second, a standard which ensures that the presentation of a sequence of completely static images appear to the human observer as smooth and continuous motion.[278] By contrast, if the entire universe flashes in and out of existence at the limiting Planck time of 5.3×10^{-44} seconds, equivalent to a "frame" rate of almost 10^{44} "frames per second," the cosmos would obviously *appear* to be smooth and continuous in all respects, certainly to any human observer of the cosmos, even at quantum dimensions.[279]

The approximate image resolution of a "holoplenum display" obtained by dividing one inch by the Planck length, yields a maximum resolution of 1.584×10^{34} holopixels per inch. At such hyper-fine resolution, even a proton in the 10^{-15} meter

---

[278] Dorf, *The Electrical Engineering Handbook*, 1538.

[279] Planck time is a constant set by the speed of light and the gravitational constant, at a value of 5.39106×10^{-44} seconds; Runehov and Oviedo, eds., *Encyclopedia of Sciences and Religions*, 1772.

dimensional range would appear to be moving smoothly through space.

Quantum Biology and the Holoflux Field

Quantum mechanics and the concepts of the holoplenum and the holoflux field provide tools for understanding the biological basis for human modes of consciousness. The anesthesiologist and consciousness researcher Stuart Hameroff has, along with the British physicist Sir Roger Penrose, proposed a quantum theory of human consciousness based upon quantum field activity within microtubule structures within the human body.[280] It is notable that in a recent paper, Hameroff argues for an approach to consciousness studies in terms of quantum mechanics, and refers to his own contribution in terms of an emerging "quantum biology:"

> At the turn of the 20th century, physics was perceived by its practitioners to be as solid as rock, only to be shaken to the core by the discovery of quantum mechanics. Today, perhaps the most dynamically expanding branch of science is molecular and cellular biology, which is amassing impressive reams of data well ahead of its proper integration and deep analysis. Quantum biology is emerging gradually as a response to the challenge of explaining such important, yet poorly understood phenomena as photosynthesis, bioenergetics, vision, olfaction, bird navigation, etc. Yet, the grandest challenge

[280] Hameroff and Penrose, "Conscious Events as Orchestrated Space–Time Selections."

> of all is to explain how our brains work and, in particular, how does conscious behavior emerge from the structure and function of the human brain and its cellular and sub-cellular components.[281]

Quantum Brain Dynamics

Quantum brain mechanics, and in particular, *Quantum Brain Dynamics*, a field pioneered by the physicists Hiroomi Umezawa and Herbert Fröhlich in the late 1960s, is one such contribution to Hameroff's proposed quantum biology.[282] As early as 1978, Stuart, Takahashi, and Umezawa proposed a mechanism of human memory and consciousness, consistent with quantum field theory, which they called "quantum brain dynamics (QBD)."[283]

Based on Umezawa's original research, Mari Jibu and Kunio Yasue at Seishin University in Japan put forward a detailed model of consciousness based upon the quantum dynamics of the electromagnetic field, which they describe as follows:

> The long-standing difficulty of understanding the origin and mechanism of unity of consciousness (as a unified self) has been called the "binding problem" in neurophysiology. The binding problem may

[281] Hameroff, Craddock, and Tuszynski, "Quantum Effects in the Understanding of Consciousness," 230.

[282] Jibu and Yasue, *Quantum Brain Dynamics and Consciousness*, 147.

[283] Stuart, Takahashi, and Umezawa, "Mixed System Brain Dynamics."

> be solved . . . by investigating the usually neglected quantum electromagnetic phenomena taking place in the dynamically ordered regions (i.e., perimembranous regions) of intracellular and extracellular water. There, each brain cell is enfolded within a common field of macroscopic condensation of evanescent photons and all the physico-chemical processes taking part in the stratified society of brain cells are subject to the control and unification by quantum electrodynamics. Unity of consciousness thus arises from the existence of the global field of condensed evanescent photons overlapping the whole brain tissue in the cranium.[284]

In extending Umezawa's theory of QBD, Jibu and Yasue describe how they have discovered that, through dynamic resonance (spin), polarized dipole water molecules throughout the blood stream produce the activity, in effect, of a single resonant water dipole, a "macromolecule" in the bodies of all living, water based creatures. The macromolecular-dipole, offering an exceptional information highway for electromagnetic fields on the macro level, manifests as a plasma. It should be noted that it was the physics of plasma that captivated David Bohm during his thesis research under Oppenheimer, and it is useful here to clarify the definition of plasma:

> A state of matter in which a significant portion of the particles are ionized. Plasmas are by far the most common phase of matter

[284] Jibu and Yasue, "Quantum Brain Dynamics and Quantum Field Theory," 288.

> in the universe, both by mass and by volume. All stars are made of plasma, and interstellar space is filled with plasma. Common forms of plasma include lightning, St. Elmo's fire, the polar aurorae, the solar wind, neon signs, and plasma displays in modern home television.[285]

Within this polarized, nonlocal, H_2O quantum field is embedded an electromagnetic field described here by Jibu and Yasue:

> There are two basic fields in QBD: the water rotational field and the electromagnetic field. The two must be described simultaneously by quantum field theory because they interact with each other . . . . The water rotational field extending over the whole assembly of brain cells manifests an overall domain structure in which the whole region occupied by water is divided into many domains . . . . In other words, they are bound to each other quite strongly by a quantum theoretical nonlocal coherence.[286]

In 1995, building upon their earlier paradigm, Jibu and Yasue published the mathematics to support their theory that the creation and annihilation dynamics of "corticons" points to an "energy quanta of the water rotational field extending to the whole assembly of brain cells, and photons, that is, energy

---

285 Chen, *Introduction to Plasma Physics*, 56.

286 Jibu and Yasue, *Quantum Brain Dynamics and Consciousness*, 164–65.

quanta of the electromagnetic field."[287] Their theory focuses upon the nonlocal quantum field that is generated within the human circulatory system, embedded in an electromagnetic field manifesting throughout the contiguous ionized , blood plasma system.

That our blood might act as an electromagnetic plasma should not be surprising— the first line in a textbook on plasma physics reads, "It has often been said that 99% of the matter in the universe is in the plasma state."[288] Jibu and Yasue go further by explaining how QBD explain how life itself can be seen to be equivalent to "the unity of water" in the human body, or as "a single molecule" in the human blood system:

> All H_2O molecules bound together in a QBD vacuum domain form a single, extensive molecule of water in a macroscopic domain. That is, water throughout the entire region of the cerebral cortex is thought to be composed of many macroscopic water molecules whose sizes are all comparable to the coherence length of the QBD vacuum, that is, about 50 microns. This remarkable feature of water in living matter might provide us with quantum field theoretical support for the idea that life is nothing but the unity of water as a single molecule in living matter.[289]

---

287 Jibu and Yasue, "Quantum Brain Dynamics and Quantum Field Theory," 293.

288 Chen, *Introduction to Plasma Physics*, 1.

Quantum brain dynamics supports the model of a locus of consciousness beyond that of the nervous system and brain. The supporting pattern can be found in the QBD manifestation of a "single extensive molecule of water," dipolar in geometry like an antenna, generating a complex magnetic field through the circulation of hydrogen ions in blood plasma.

In such a model, the blood system of the human body can be seen as an extensively polarized "super cell" of nonlocal electromagnetic plasma energy, which can then be differentiated from the neuronal brain body of consciousness, itself generated by sequential electrical patterns flowing as pulses in the nervous system pathways. Moving charges generate magnetic fields, and ionized human blood flow is no exception: flowing blood plasma results in creation of a magnetic field, and this is in accord the conjecture of QBD.[290] The circulatory system can be seen as a magnetic plasma composed primarily of ionized red blood cells (referred to as erythrocytes) and water molecules, flowing together in complex vortices of blood plasma around every cell and through every capillary of the body.[291] Each erythrocyte is a flexible, annular, bio-concave disk shaped like a doughnut (in geometry, a torus), having a thin webbed center where the hole in a pastry doughnut would be located (Fig. 34).

---

[289] Jibu and Yasue, "Quantum Brain Dynamics and Quantum Field Theory," 293.

[290] Dorf, *The Electrical Engineering Handbook*, 27.

[291] McCraty, *The Energetic Heart*.

Mammalian erythrocytes are unique among vertebrates for having no nucleus at all; each blood cell sheds its nucleus and internal organelles as it reaches maturity. The typical outside diameter of a red blood cell is approximately 9 microns, close to the infrared wavelength of 9.6 microns generated by the human body.[292]

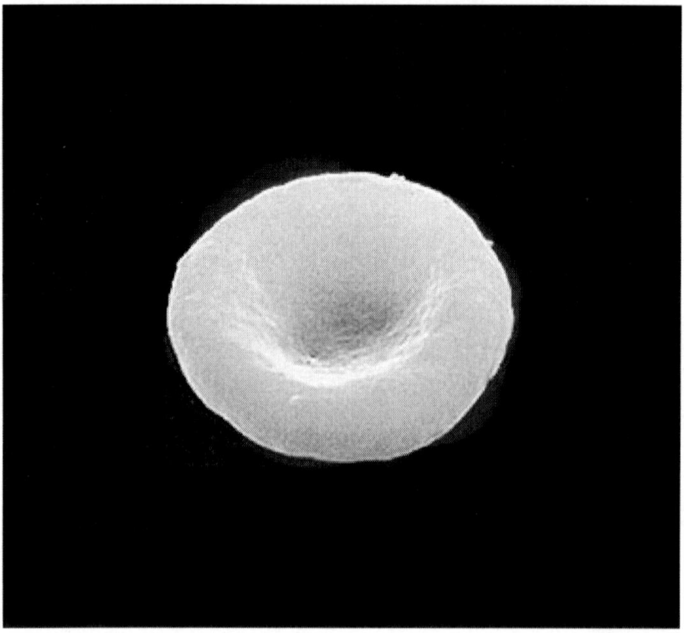

Figure 26. Red blood cell. Graphic by Svdmolen (2005). This work is in the public domain under the terms of Title 17, Chapter 1, Section 105 of the US Code.

The adult human body contains approximately 6 grams of iron, of which 60% is stored throughout the 10^{12} erythrocytes, each of which contains approximately 270 million atoms of ionic iron

---

292 Turgeon, *Clinical Hematology*.

embedded in transparent hemoglobin.[293] Thus each erythrocyte, replete with iron ions embedded in hemoglobin, creates in effect, an ionized iron toroid.[294] One of the best-known applications of the toroid shape is called the "Tokamak device," which uses a magnetic field to confine plasma in the shape of a torus in experiments to produce controlled thermonuclear fusion power. Andrei Sakharov, who eventually became known internationally as a Soviet dissident, first designed the Tokamak enclosure device in the Soviet Union in the 1950s.[295]

Holoflux Analysis of Human Red Blood Cell

Let us step through a holoflux analysis of the possible functions of a red blood cell as a locus of consciousness and the possible use of the erythrocyte as a locus of memory storage at the human biological scale. If, as previously conjectured, the red blood cell has an ideal diameter to resonate electromagnetic radiation in its ferrite-embedded ring at the human infrared wavelength of 10 microns, then we can ask if this configuration could accommodate a unique isospheric frequency (wavelength) for each of the currently 7 billion living humans on the planet. Figure 35 outlines the feasibility of this approach.

Assuming each unique frequency would match its radially unique isosphere, separated by only one

[293] Romanes, *Cunningham's Textbook of Anatomy.*

[294] Wick, Pinggera, and Lehmann, *Clinical Aspects and Laboratory Iron Metabolism*, 6.

[295] Nishikawa and Wakatani, *Plasma Physics.*

Planck length, Figure 35 suggests how seven billion unique isospheres, each of quantum discrete frequency, might be nested within the geometry of a typical human red blood cell (in the image we have multiplied 7 billion times the Planck length of approximately 10^{-35} m to obtain an estimated shell thickness of 10^{-26} meters). This implies that each living human being might have a unique holospheric frequency, detectable by other human blood cells via the implicate order about which each is centered, and thus provides a possible mechanism for communication, via the mechanisms of resonance and superposition in the frequency domain of the implicate order.

How 7 billion unique human frequencies can be accommodated within the ring diameter of an erythrocyte.

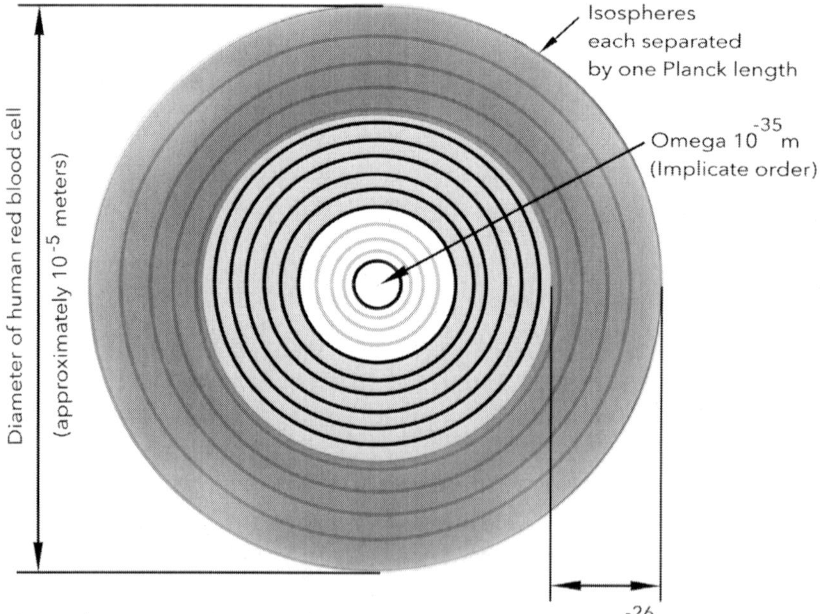

7 billion human isospheres separated by 1 Planck Length = 10^{-26} meters
(not to scale)

Figure 27. Isospheric capacity of red blood cell. Graphic by author.

A calculated estimate of the memory storage capacity of a single red blood cell can now be made, using Wheeler's approach to determine the number of Bekenstein equivalent bits on the surface of an isosphere the diameter of a typical red blood cell.[296] Here we take 8.1 microns, (or 8.1 x 10^{-5} meters), as the average diameter of a typical human red blood cell, and proceed to calculate the maximum possible storage capacity on the surface of a hypothetical holosphere associated with a single red blood.[297] To obtain the limiting number of bits, we simply calculate the surface area on a sphere of 8.1 microns diameter, and then divide the value obtained by 1.04 x 10^{-69} square meters (which is the Bekenstein unit of entropy, or approximately the square of the Planck length, 1.616199 x 10^{-35} meter).[298] The surface area of the erythrocyte-bisected sphere is $4\pi r^2$or $4\pi(8.1 \times 10^{-5})^2 = 4\pi(6.561 \times 10^{-9}) = 8.24 \times 10^{-9}$ square meters. Dividing this by the qubit area of 1.04 x 10^{-69} square meters yields the estimated maximum storage capacity of 8 x 10^{60} information bits. This is an extremely large information storage capacity.

Not to be neglected as part of any electromagnetic holoflux process associated with the cardiovascular system is the likely role of the heart:

---

[296] Wheeler, *A Journey into Gravity and Spacetime*, 222.

[297] Romanes, *Cunningham's Textbook of Anatomy*, 137.

[298] Wheeler, *A Journey into Gravity and Spacetime*, 222.

> The heart generates the largest electromagnetic field in the body. The electrical field as measured in an electrocardiogram (ECG) is about 60 times greater in amplitude than the brain waves recorded in an electroencephalogram (EEG). The magnetic component of the heart's field, which is around 5000 times stronger than that produced by the brain, is not impeded by tissues and can be measured several feet away from the body with Superconducting Quantum Interference Device (SQUID) based magnetometers.[299]

Dimensional Analysis and EMF Wavelength

As previously discussed, a rule of thumb in applying the phenomenon of resonance in any situation indicates that the optimal physical dimensions of the resonant receiving system must approach a half-wavelength of the dimension of the transmitted energy wave. Using dimensional analysis, this fact can be used in the search for possible loci of consciousness in the electromagnetic spectrum. In this section we examine calculated correlations between dimensions of structural bio-components in the human body and corresponding wavelengths in the electromagnetic spectrum.

If we examine human biophysical components and calculate their peak resonant electromagnetic energy frequencies using Wien's Law, and then identify the frequency bandwidth within which we find these resonances for the biological components being

[299] McCraty, *The Energetic Heart*, 1.

examined, two possible systems emerge: the near infrared and the near ultraviolet (the use of the word "near" implies frequencies adjacent to the visible spectrum).

The following dimensional analysis correlations support the possibility of a conscious electromagnetic–holoflux information system active within a 10 micron wavelength band throughout the human bio-system:

- The human body at 98.6 F emits wavelengths of 9.6 μm.
- Earth has an atmospheric pass-band at wavelengths of 9–10 μm.
- Inner diameters of blood capillaries are 10 μm.
- Outside diameter of a typical blood cells is 9 μm.
- CO_2 laser outputs between wavelengths of 9.4 –10.6 μm.[300]

Data is presented in Table 5, ranking physical components by increasing size, along with their calculated Wien blackbody peak wavelengths and corresponding theoretical waveguide resonant frequency ranges.

As indicated in Table 5, the average diameter of human blood capillaries of 10 μm coincides with a major "passband" of the planetary atmosphere itself. As previously discussed, a simple Wien's law

[300] The carbon dioxide laser is one of the earliest gas lasers to be developed (invented by Kumar Patel of Bell Labs in 1964), and is still one of the most useful. Carbon dioxide lasers are the highest-power continuous wave lasers that are currently available; Milonni and Eberly, *Laser Physics*.

calculation show that the normal human body radiates most strongly in the near infrared, measured at a macro level as "heat" at a temperature of 91.6 F, which again, using Wien's Law, exhibits a peak spectral energy response at wavelengths of approximately 10 μm. It is accordingly to this wavelength that the maximum sensitivity of military night-vision goggles is tuned.[301]

A physiological waveguide candidate in another frequency range can be found in the inner diameter of the microtubule system, located throughout neurons and also within cilia.[302] At 15 nm in dimension, microtubules would be especially sensitive to an electromagnetic wavelength in the ultraviolet bandwidth, just above visible light frequencies in the electromagnetic spectrum.

Table 5
Physical Correlations with Electromagnetic Frequency Bands

---

[301] Clark, *Photography by Infrared—Its Principles and Applications*, 78.

[302] Pizzi et al., "Evidences of New Biophysical Properties of Microtubules."

Biosphere, Magnetosphere, Chromosphere

In searching for energies of consciousness, it is important to explore the possibility of consciousness as having a component (if not comprising) the Earth's magnetic field, also known as the geomagnetic field, the magnetic field of energy flux extending from the Earth's inner core out to its interaction with that stream of high energy particles emanating from the Sun, the solar wind. This enormous magnetic field in which we find ourselves is generated by a magnetic dipole, powerfully driven by a spiraling ocean of liquid metal in the outer core, spinning around the enormous copper-magnesium crystal of the Earth's inner core. A supercomputer at the Los Alamos National Laboratory has created a model of flow patterns of geomagnetic energy within and around the Earth, illustrated in Fig. 36. Each line represents the locus of an intense magnetic field component. The dense clusters of magnetic lines are within the Earth's core.

Figure 28. Supercomputer model of magnetosphere. Graphic by Glatzmaier (2007). This work is in the public domain under the terms of Title 17, Chapter 1, Section 105 of the US Code.

The dipole axis of the Earth is tilted at an approximate angle of 11 degrees with respect to the rotational axis, as if there were a bar magnet placed at that angle along the central axis of the Earth. However, unlike the static field of a bar magnet, Earth's geomagnetic field quivers and resonates over time in synch with the motion of molten iron alloys in the Earth's outer core, the geodynamo. If the universe is alive and sentient, then these vibrations could be replete with frequency modes of consciousness, communication signals yet to be deciphered by human technology, filling bandwidths that have perhaps only thus far been explored by mystics, shamans, and cybernauts.[303]

Where might there be found other electromagnetic energies of the earth, candidates for some holoflux-driven form of consciousness? Geophysicists have determined that two fundamental energy sources are to be found in the earth's crust: geothermal and geoelectric energy, respectively. We are all aware of geothermal energy, spectacularly displayed in enormous glowing lava flows from volcanic activity, erupting geothermal energy rising from below the crust of the Earth. But what is not widely known is that while the amount of *geothermal* energy in the world is estimated to be approximately equal to 2×10^7 Megawatt-hours (MWh), this is dwarfed by the amount of *geoelectric* energy, which is equal to 1.14×10^8 MWh, a surprisingly tenfold magnitude greater than the geothermal energy.[304]

---

303 Lilly, *The Deep Self: Consciousness Exploration in the Isolation Tank.*

304 Booth, Koren, and Persinger, "Increased Feelings of the Sensed Presence During Exposures to Weak Magnetic Fields."

This geoelectric energy, which can be seen in the atmosphere discharging in lightning strikes and in ionized aurora borealis displays, flows in vast currents of electromagnetic energy, rivers of flowing electrons, both in the sea and within the Earth's crust, in enormous rivers of electricity.[305] These geoelectric flows are called telluric currents (from Latin *tellūs*, "earth"), and they occur both in the Earth's crust as well as within the much deeper mantle. These geomagnetically induced currents are generated by the daily enormous magnetic interactions of the sun's magnetic field and the spinning Earth's geomagnetic field at the geopause, the area where the sun and the earth meet in ever-changing resonant magnetic vibrations of enormous power that induce the aurora borealis and the telluric currents flowing in and around the surface of the planet.[306]

Nikola Tesla, who invented the alternating current electrical grid system, in spite of Thomas Edison's attempt to make his direct current distribution system the standard, began to fabricate a system that would tap telluric currents for electrical power and use the currents themselves to transmit the power in wireless electrical energy grids, affording free power distributed over the face of the Earth by the planet itself.[307] Mysteriously, his pilot experimental station which would have tapped the Telluric currents

---

[305] Allaby and Allaby, "Telluric Current."

[306] Ibid.

[307] M. Cheney, *Tesla: Man Out of Time.*

was burned to the ground, some say by Thomas Edison's supporters.[308]

This enormous geoelectric energy is spread all over the earth's crust, most concentrated in wet rocks located in the earthquake belt zones.[309] If human consciousness does have an energy component in the electromagnetic spectrum, then it must be affected by, and interact to some degree with, this geoelectric energy environment of the earth. Many animals possess a "magnetic sense" and appear to be able to track the magnetic field of the earth, which is generated by the flow of molten material in the earth's core and the corresponding flow of ions in the atmosphere.[310] Magnetoreception is an accepted phenomenon among a wide range of animals: birds, fruit flies, honeybees, turtles, lobsters, sharks, stingrays, whales and even bacteria.[311]

Even animals not normally known for their migration habits have been discovered to possess such a sense:

> Recent publications from a German research group (Begall et al., 2008), made the discovery that cattle (and other herd animals, such as red and roe deer) tend to

[308] Ibid, 78.

[309] Vejvoda, *17th International Conference on Reactor Technology*, 17.

[310] Ruhenstroth-Bauer, "Influence of the Earth's Magnetic Field," 336.

[311] Wiltschko and Wiltschko, "Magnetoreception in Birds and Other Animals," 675.

> situate themselves on a magnetic North–South axis, as if involuntarily directed by the earth's magnetic field. These surprising results were discovered when satellite images provided by Google Earth were used to analyze herding patterns and behavior. However, the built-in magnetic compass gets out of alignment the closer the cattle get to high voltage power lines, and the cattle then align with the power lines instead.[312]

Contemporary research in electrophysiology indicates that human bodies may be more involved in sensing electromagnetic fields than has previously been acknowledged. Research at Cal Tech has discovered traces of magnetite in the human brain and heart, in about the same density as that found in migrating animals, and has proven that onset of rapid eye movement in sleeping humans is shortened in the E–W directional orientation of sleepers as compared to the N–S position.[313]

On the average there are from about one hundred lightning discharges per second around the globe.[314] It is known that one of the physical characteristics of the earth is that an effective waveguide cavity exists between the ionosphere and the surface of the earth, and this cavity resonates at 7.8 to 8.0 Hz, in what is called the Schumann resonance

---

312 Ibid, 677.

313 Ruhenstroth-Bauer, "Influence of Earth's Magnetic Field," 195.

314 Persinger, "Schumann Resonance Frequencies," 25.

frequency peak.[315] The same frequency peak found in high end Theta waves and low end Alpha waves as measured on the human scalp.[316] In studying this Schumann cavity resonance phenomenon, the neurophysiologist Michael Persinger hs concluded:

> Because of the very small attenuation rates within the cavity, electromagnetic patterns are generated for megameters without appreciable loss of amplitude. In general, the expansion and return of the "ringing" around the earth's circumference can occur between 4 and 5 times. The latency for the return of the circular wave is about 125 milliseconds.[317]

These lightning discharges, within the geophysical cavity, emit energy waves that travel around the globe and return in 125 ms cycles. In one second, each wave circles the planet eight times (8 x 125 ms = 1 second). This in effect results in an ongoing lightning driven pulse rate equal to the cranial wave frequency of 8 Hz. The question then becomes, is this match of the predominant human Alpha wave with the planetary Schumann resonances of electromagnetic energy coincidence, or evolutionary design? If the psyche has a spectral frequency component, then is it possible that there have evolved in the human biosystem physiological modulators of this frequency component?

[315] Sentman, "Schumann Resonances," 271.

[316] Allaby and Allaby, "Telluric Current," 43.

[317] Persinger, "Schumann Resonance Frequencies," 25.

Microwave cavities are to the invisible frequency energy spectrum what lenses are to visible light energy.[318] In a biomedical engineering class lecture discussing the electrophysiology of the nervous system, I noticed the striking similarities between the waveguide horn antennas designed in another class, a microwave communication lab, and the images of the ventricular cavities within the human cranial cavity. I mentioned this to my professor and was assured that the ventricular system, as understood, had only two functions, that it existed to help maintain thermal equilibrium in the brain, and that it acted as mechanical protection for the brain, by attenuating sudden shock trauma to the skull. Nevertheless, the ventricular cavities (Fig. 37) are seen to be dimensionally consistent with the microwave band, which covers a range of frequencies with wavelength from one millimeter to one meter.[319]

---

[318] Warren and Brandt, "Optical Constants of Ice from the Ultraviolet to the Microwave."

[319] Dorf, *The Electrical Engineering Handbook.*

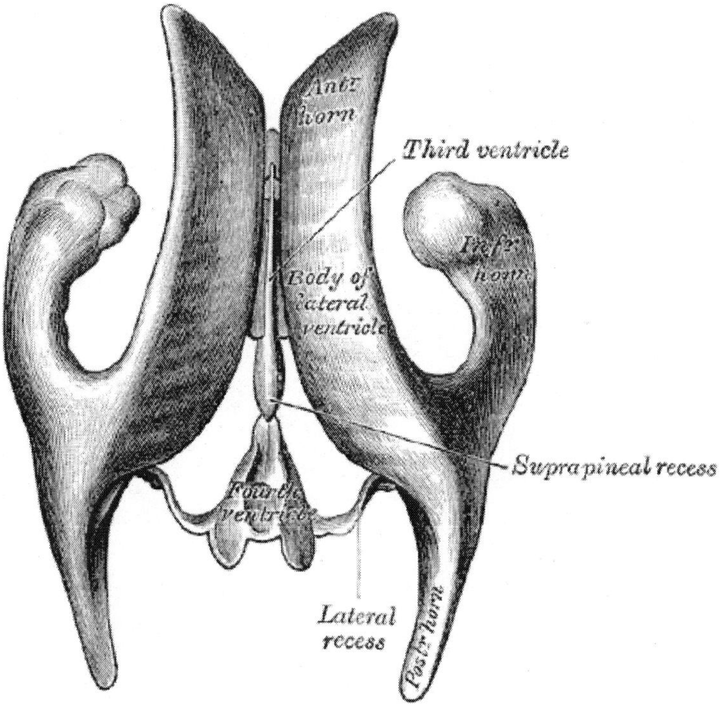

Figure 29. Drawing of ventricular cavities. Graphic by Gray (1918). Reprinted under the terms of a Creative Commons Attribution ShareAlike 3.0 Unported license. Image retrieved from Wikimedia Commons.

Another clue may be seen in cerebrospinal fluid itself, conventionally dismissed as being understood simply to act as a shock absorber or nutrient circulation system for the brain:

> Cerebrospinal fluid (CSF) is a clear, colorless liquid that bathes the brain and spinal cord. While the primary function of CSF is to cushion the brain within the skull and serve as a shock absorber for the central nervous system, CSF also circulates nutrients and chemicals filtered from the

blood and removes waste products from the brain.[320]

Comparing the drawing of the ventricular cavities (Fig. 37) with the image of a waveguide system for microwaves used in radar (Fig. 38), it can be concluded that perhaps within the cavities of the ventricular system can be discovered another instance of Deacon's "something that is not there."[321]

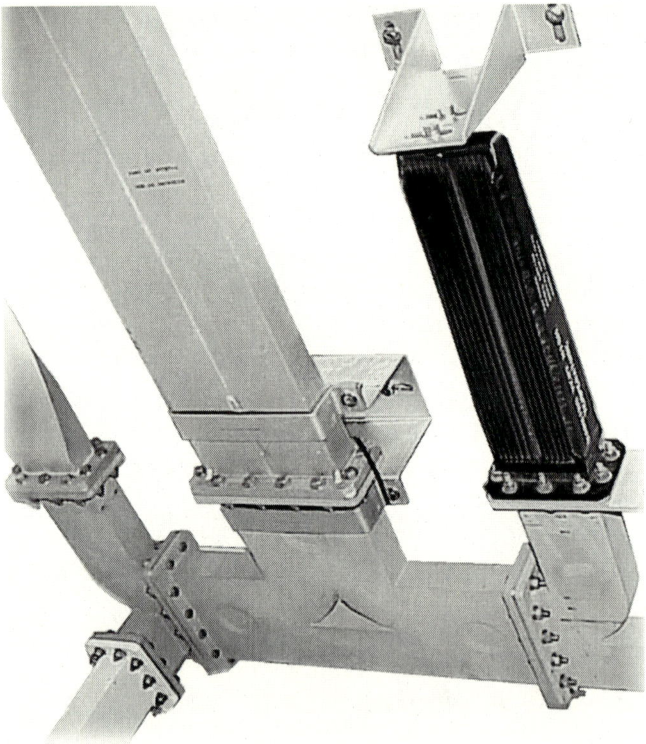

---

320 National Multiple Sclerosis Society. "Cerebrospinal Fluid (CSF)."

321 Deacon, "What Is Missing from Theories of Information," 140.

Figure *30*. Microwave radar circuitry. Graphic by Averse (2007). Reprinted under the terms of a Creative Commons Attribution ShareAlike 3.0 Unported license. Image retrieved from Wikimedia Commons.

Chapter 4: Summary and Conclusions

This book has presented a theory that the ground of consciousness, consciousness itself, manifests as a continuously-transforming flow of energy in a two-way inward/outward flow between two domains whose boundaries are omni-present. This energy flux is encoded with both information and meaning, flowing alternately between the explicate domain of space–time and the nonspatial, nontemporal transcendent domain which David Bohm termed the implicate order, a domain that exists within the center, everywhere, of space–time. To express the interconnectedness of this energy flux, spanning a wholeness that embraces both space-time phenomena as well as transcendent phenomena, we have selected the term *holoflux*, from the root term holism, defined here:

> **Holoflux:**
> The fundamental interconnectedness of all phenomena, and within this fundamental interconnectedness not only is each entity, relationship, experience and phenomenon a hollow or whole, in its own right, but all wholes are held together by a great unifying force.[322]

---

[322] Collister, *A Journey in Search of Wholeness and Meaning*, 69.

The Holoflux Theory of Consciousness

The holoflux theory of consciousness has been constructed upon these underlying concepts from Pribram's holonomic brain theory and Bohm's ontological interpretation of quantum theory.

These theories have been discussed here and shown to be congruent, supported and knit together by established principles of electrical engineering. The holoflux theory proposed here can be categorized as a cosmological ontological process structured as follows:

- A CENTERED COSMOS: In agreement with the theory of general relativity, the holoflux theory is founded upon the notion that the universe has a center, and that center is everywhere.

- HOLOFLUX: Dark energy of the implicate order, outside of space–time; Karl Pribram's theorized flux within David Bohm's implicate order; it is an energy of sentient awareness that views the outwardly explicate world of space–time from within an implicate order.

- QUANTUM BLACK HOLES: Quantum black holes, *Planck holospheres* of the Bohmian implicate order; bounded by spherical shells of Planck length diameter 10^{-35} meters.

- THE HOLOPLENUM: An invariant *holoplenum* of close-packed holospheres, or *isospheres*, underlying and from which is projected the explicate order; the holoplenum fills a continuum of isospheres everywhere in the physical universe from the center of each position in space–time; the concept had been intuited by Leibniz in his theory of the monads.

- RESONANCE: Holoflux frequency resonance connects the explicate order (the outside world), with the implicate order (the inside world); holoflux communication occurs everywhere between the implicate order (frequency-phase flux in the implicate order, f_i) and resonating isospheres of the explicate order in spherical shells. These holo-shells must be separated at corresponding distances, λ_i, from each central Planck holosphere, to maintain the relationship, $f_i\ \lambda_i$,= c, where c is the speed of light in a vacuum, $3x10^8$ m/sec.

- CYCLIC PROCESS: A flowing movement, an *electromagnetic-holoflux* movement of energy continually cycles between the implicate order and the explicate order; the flow can be mathematically modeled through Fourier transforms and the Bohm quantum potential wavefunctions.

- A COSMIC CLOCK: Granular units of Planck time (t_P), vibrate at a "clock-cycle" rate of 5.39106×10^{44} cycles per second synchronizing the flow between alternating manifestations of implicate and explicate frequency information; in Cartesian coordinates, this can be pictured as a square wave, a Planck-time clock.

- BOHM'S QUANTUM POTENTIAL FUNCTION: Conceptional computations within the implicate order are "expressed" in space–time (the explicate order) via the effect of Bohm's quantum potential, *Q*.

- A FOURIER LENS: A Fourier-type mathematical transform can be seen to act as a lens through which the implicate "views" or "cognizes" the explicate order of space–time; this Fourier lensing is the process by which the implicate order retrieves, observes, and experiences information generated within the explicate order.

A diagram of these assumptions is presented in Fig. 50, "Solving the Hard Problem of Consciousness," in which the relative positions of essential concepts from the following philosophers have also been indicated:

- David Chalmers' "structure of awareness" is posited to be synonymous with the implicate order, and his "structure of consciousness" accordingly is shown within the explicate order.[323]

- Ned Block's "A-consciousness" lies in the implicate order, and his "B-consciousness" in the explicate order.[324]

- David Bohm's "meaning" lies in the implicate order, and his "information" in the explicate order.[325]

---

[323] Chalmers, "Facing Up to the Problem of Consciousness."

[324] Block, *Consciousness, Function, and Representation.*

- László's "A-dimension" lies in the implicate order, and his "B-dimension" lies in the explicate order.[326]

The process can be visualized as shell-like waves of radiant information-embedded energy spreading out and intersecting with the myriad of the other expanding shells of active information from all of the other "particle/holospheres" in the holoplenum-filled cosmos. The intersection of these waves and their condensation, where the density of intersection is maximum, is regarded in classical physics as "matter," but this can also be seen as a process, an on-going holographic extrusion of information *from* the implicate order *into* the explicate order, where the manifestations in space–time are holographically projected.

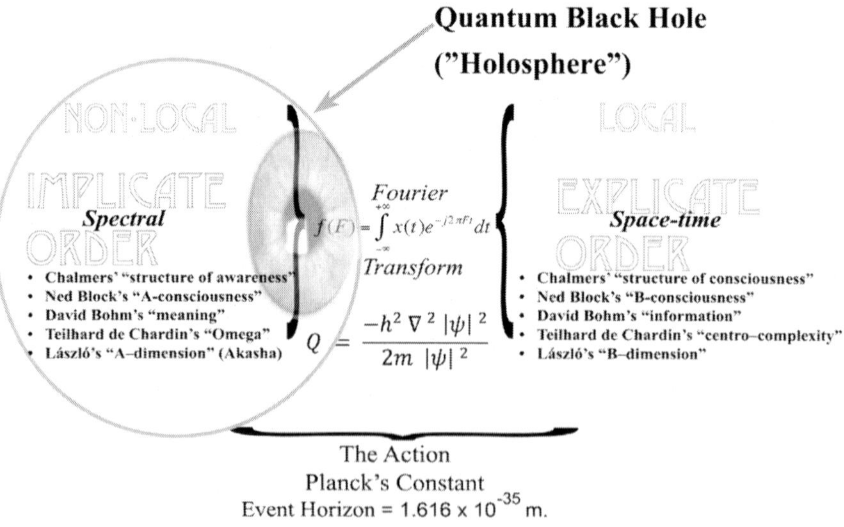

---

[325] Bohm, *Wholeness and the Implicate Order*.

[326] László , *The Self–Actualizing Cosmos*.

Figure 50. Solving the hard problem of consciousness.

In the diagram, an iris-like lens at the Planck length aperture, between David Bohm's implicate order and the explicate order, mirrors Karl Pribram's 1991 conceptualization in *Brain and Perception*:

> These two domains characterize the input to and output from a lens that performs a Fourier transform. On one side of the transform lies the space–time order we ordinarily perceive. On the other side lies a distributed enfolded holographic-like order referred to as the frequency or spectral domain.[327]

Note that the image of an iris in fig. 50 appears at the edge of the event horizon of a quantum black hole, or implicate order holosphere. The iris symbolizes consciousness looking out from the implicate order into space–time via a Fourier transform lensing process. This can also be understood as the observer observing the observed, as in the context of Wheeler's "Self-observing U."[328] Two equations can be seen in the center of the figure, depicting Pribram's holonomic Fourier transform, and Bohm's quantum potential function *Q*. These illustrate the ongoing mathematical lensing-flow of conscious energy between the local (explicate) and nonlocal (implicate) regions of reality.

---

[327] Pribram, *Brain and Perception*, 70.

[328] Wheeler, *Geons, Black Holes, and Quantum Foam*, 339; also see fig. 21.

In projecting the cosmos, nonlocal holoflux energy in the spectral domain resonates actively with electromagnetic plasma in the space–time domain; thus the nonlocal affects the local, explicate order. This information emerges as electromagnetic energy at the Planck wavelength, moving outwardly from the implicate center in spherical waves of electromagnetic energy, into the explicate.

The flow of holoflux energy into space–time can also be expressed in electrical engineering imagery, topologically, as a variation of Ampere's "left-hand rule." [329] The three domains can be visualized with the thumb, index finger, and middle finger of the left hand held mutually perpendicular to indicate the three orthogonal fields: the electric domain (e_d), the magnetic domain (m_d), and the frequency flux domain (f_d). We hypothesize that all three domains have equally valid ontological reality, but that only two of them, the magnetic and the electric, manifest explicitly in the space–time domain. Conversely, the third domain, the frequency flux domain (f_d), manifests "within" the implicate domain, transcendent of space and time.

We are drawn to consider a magnetic system of consciousness in the locus of the cardiovascular system. The entire system of magnetically polarized molecules in the bloodstream, throughout the more than 25,000 miles of veins and capillaries, might act as a fully connected waveguide system at infrared light frequencies.[330] A parallel can be seen in twenty-first-

---

[329] Nilson and Hornung, *Practical Radio Communication,* 54.

century internet technology, where glass fiber optic cables interconnected over hundreds of thousands of miles act as waveguides for networks at visible light frequencies. Thus an entire system, a pathway of information-modulated consciousness, might exist, in addition to the electrical nervous system distributed throughout the brain and body in a neuronal electrical network.

Strategies for Testing the Holoflux Hypothesis

If consciousness is manifesting as energy flow in electromagnetic-frequency fields, then one should be able to determine experimentally the location of high bandwidth network channels within human physiology, channels by which electromagnetic information interchange guides the growth process, effects repair, and catalyzes evolutionary mutation. The widespread assumption of material science has been that consciousness emerges from neuron activity in the human brain only as an epiphenomenon.[331] Accordingly, where then might a neurophysicist search, outside of neuronal linkages?

Dimensional evidence indicates that good candidate ranges for testing the electromagnetic field component of consciousness can be found in the infrared spectrum. This region lies just below the threshold of the visible spectrum that is picked up by eye cone structures.[332] Human core body

---

[330] Zimmer, "Origins: On the Origin of Eukaryotes," 667.

[331] Dennett, *Consciousness Explained*, 406.

temperatures, ranging from 36.3 C to 37.5 C on a diurnal cycle, indicate that, applying Wien's Law, we should search the infrared spectrum in a bandwidth between 9.36425 microns and 9.32808 microns, near the middle of the infrared.

The range of human body temperature associated with human consciousness spans an exceptionally wide frequency band. Fig. 51 displays this range as spanning 9.88 to 9.11 microns, charted as corresponding wavelength.

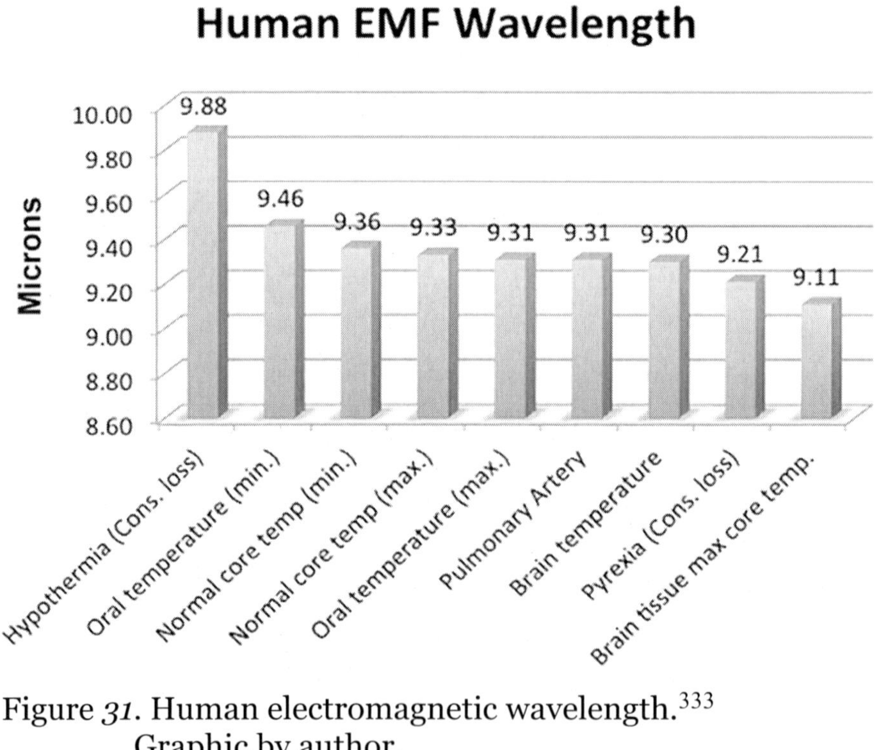

Figure *31*. Human electromagnetic wavelength.[333] Graphic by author.

---

[332] Oyster, *The Human Eye: Structure and Function.*

[333] Data from Geiger, Aron, and Todhunter, *The Climate Near the Ground*, 452.

Human body temperature ranges from a minimum of 68.0 F (hypothermia with loss of consciousness) to a maximum of 106.7 F (maximum core brain temperature before death).

Converting from wavelength, this range is equivalent to a frequency range of 30.3 gigahertz to 32.0 gigahertz, an enormously wide band compared to, for example, the FM radio frequency band, ranging from 87.0 to 88.1 MHz. Over 1000 equivalent FM radio stations could be easily broadcast within the human infrared radiation band, with no overlapping interference.[334]

Assuming that electromagnetic energy in the infrared range networks consciousness within the human body, where might we search to detect such activity? One approach in the search for an infrared component of consciousness would be to monitor the dynamics of an infrared spectrum emanating from and external to the human body in an attempt to detect information laden photons escaping the body as modulated infrared radiation.

> Interestingly, because biologic materials are transparent to light in the near-infrared region of the light spectrum, transmission of photons through organs is possible.[335]

---

[334] Calculations in this paragraph use Wien's Law to convert wavelength to frequency.

[335] Cohn, *Near-Infrared Spectroscopy: Potential Clinical Benefits in Surgery*, 323.

Another approach would be to search for infrared energy signals flowing as patterns within physiological waveguide channels within the human body. The ubiquitous blood capillary system, for example, with typical inner diameters of 10 microns, is a likely candidate to be acting as an electromagnetic waveguide. Capillaries provide a ready-made network infrastructure within which the flow and resonance of a modulated infrared energy plasma might be discovered.

Other bandwidths than the infrared should not be ignored in the search for an electromagnetic energy mode of conscious activity. If we consider the much smaller microtubules which are embedded within the neurons, we find them to be excellent candidates to act as waveguides for electromagnetic frequency propagation at an even higher frequency band. In July 2015, Stuart Hameroff provided the following update on his research in the area of microtubules and anesthesiology, and new support for Pribram's holonomic brain/mind theory:

> Actually, the brain is looking more like an orchestra, a multi-scalar vibrational resonance system, than a computer. Brain information patterns repeat over spatiotemporal scales in fractal-like, nested hierarchies of neuronal networks, with resonances and interference beats. . . .Holograms encode distributed information as multi-scalar interference of coherent vibrations, e.g. from lasers. Pribram lacked a proper coherent source, a laser in the brain, but evidence now points to structures inside brain neurons called microtubules as sources of laser-like coherence for the

> brain's vibrational hierarchy? . . .microtubules have quantum resonances in gigahertz, megahertz and kilohertz frequency ranges (the work of Anirban Bandyopadhyay and colleagues at National Institute of Material Science in Tsukuba, Japan). . . .These coherent "fractal frequencies" in microtubules apparently couple to even faster, smaller-scale terahertz vibrations among intra-tubulin "pi electron resonance clouds," and to slower ones, e.g. by interference "beats" giving rise to larger scale EEG.[336]

Applying Wien's Law to the inner diameter of a typical microtubule results in a wavelength in the ultraviolet band, quite a bit higher in frequency than the infrared, but just on the other side of the "visible spectrum" band used by eye rods and cones during normal daylight visual reception. In 2010, a team exploring microtubule resonance at the University of Pavia's Nano-Biotechnology laboratory verified the existence of electromagnetic resonance of microtubules at a frequency of approximately 1510 MHz.[337] The team concluded the following:

> The analysis of the results of bi-refringence experiment highlights that the microtubules react to electromagnetic fields . . . at a frequency of 1510 MHz . . . . Considering the nanoscopic size of microtubules, the

---

[336] Hameroff, "Is Your Brain Really a Computer, or Is It a Quantum Orchestra?", para. 5, 6, 8, 9.

[337] Pizzi et al., "Evidence of New Biophysical Properties of Microtubules," 1.

> resonance analysis would be more effective if carried out on much higher frequencies (up to 100 GHz), with suitable instrumentation. The presence of a small but sharp resonance effect at a low frequency could be the hint of a much more evident effect at even higher frequencies.[338]

Confirmation of the generation of such electromagnetic frequencies is described in a similar study completed in the following year at another research laboratory as reported in the *Journal of Theoretical Biology*, which describes additional evidence for high frequency electromagnetic field oscillations associated with microtubules:

> The electromagnetic radiation and field characteristics of the whole cellular microtubule network have not been theoretically analyzed previously. The results indicate that macroscopic detection system (antenna) are not suitable for measurement of cellular electrodynamic activity in the radiofrequency region since the radiation of single cells is very low (lower than 10^{-20} Watts). Low noise nanoscopic detection methods with high spatial resolution, enabling measurements in the cell vicinity, would be desirable, in order to measure cellular electrodynamic activity reliably.[339]

---

338 Ibid., 13.

339 Havelka et al., "High-Frequency Electric Field and Radiation Characteristics of Cellular Microtubule Network," 31.

Some interest has already been shown in the function of microtubules as antennas, for example in ciliogenesis. Cilia are present on almost every cell type in the human body and are composed of a microtubule-based core structure, that have been seen to "act as antennae" to sense extracellular signals, such as hormones, odorants, and changes in the local environment.[340]

Several possible ranges for electromagnetic networks of consciousness are indicated in Fig. 52, "Eight Feasible Bands for Consciousness." Note that while microtubules are 40 nm in diameter, even smaller is the average 20 nanometers synaptic cleft width, which would make the interstitial space between synapses resonant in the ultraviolet range.

---

[340] Ishikawa and Marshall, "Ciliogenesis: Building the Cell's Antenna," 222.

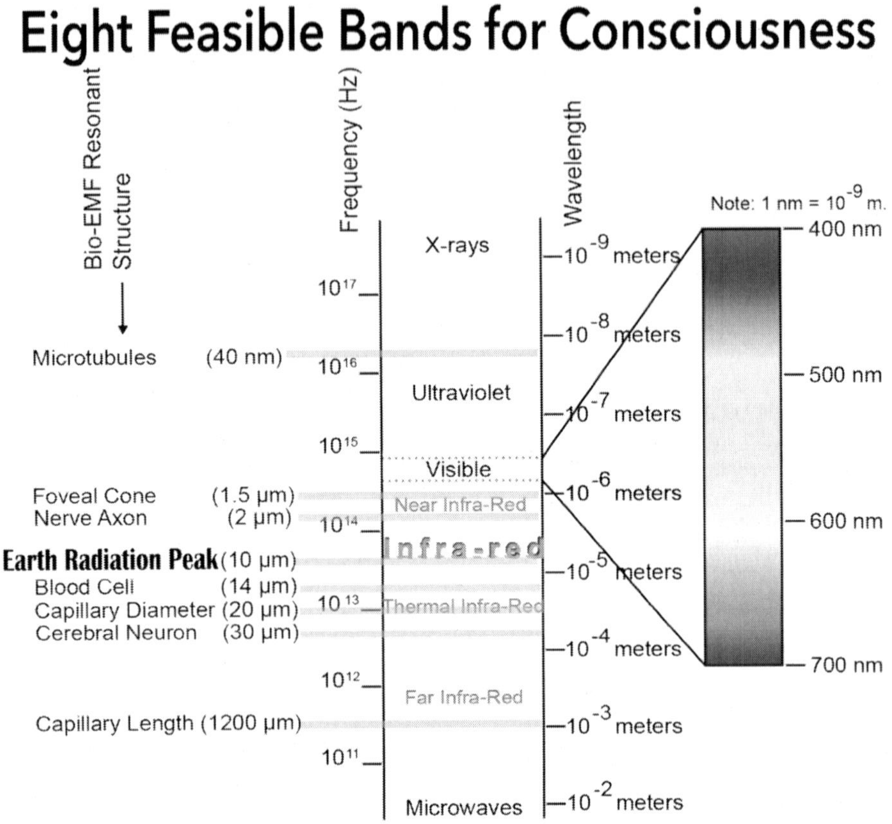

Figure 32. Eight feasible bands for consciousness. Annotations added by author to graphic by Jahoe (2012). Reprinted under the terms of a Creative Commons Attribution ShareAlike 2.5 Generic license. Image retrieved from Wikimedia Commons.

The adult human brain is estimated to contain "approximately 20 billion neocortical neurons, with an average of 7,000 synaptic connections each," making these structures possible candidates for involvement in processing and memory storage; every cubic centimeter of cerebral cortex contains roughly a trillion synapses.[341]

[341] Drachman, "Do We Have Brain to Spare?", 1.

To test whether consciousness modulates electromagnetic field activity, we should begin with the monitoring, recording, and frequency domain analysis of electromagnetic spectra radiating from biological structures. We should also measure, record, and analyze what has previously been regarded only as signal "noise" in the infrared energy radiating through the blood capillary system.

Further analysis should search for modulated patterns encoded within these spectral recordings that might indicate that they carry information. This would be similar to the research methodology conducted by John Lilly in the 1950s, in a research project funded by the National Science Foundation during which Lilly's team monitored and recorded audible dolphin/human communications, and applied Fourier frequency domain analysis in the search for super audible information patterns.[342]

Uploading Consciousness from Wetware to Hardware

In conclusion, it is possible to speculate on practical implications of this holoflux theory for future technologies. Identification of an electromagnetic energy field manifesting information characteristics in biological systems would support the feasibility that, in principle, major components of consciousness might eventually be stored, maintained, and manipulated using hardware (e.g., fiber optics and silicon) instead of wetware (biological tissue). An intriguing

[342] Lilly, *The Mind of the Dolphin: A Nonhuman Intelligence.*

implication would be that consciousness might be uploaded from wetware into such a hardware environment, and vice versa.

REFERENCES

Aetheling. 2012. "File:World-Population-1800-2100.svg" (graphic file). November 22, 2015. Wikimedia Commons. Retrieved from https://commons.wikimedia.org/wiki/File:World-Population-1800-2100.svg.

Aigner, Dennis J. 2000. *The Swastika Symbol in Navajo Textiles*. Laguna Beach, California: DAI Press.

Alabiso, Carlo, and Ittay Weiss. 2015. *A Primer on Hilbert Space Theory: Linear Spaces, Topological Spaces, Metric Spaces, Normed Spaces, and Topological Groups*. New York: Springer.

Alkari. 2012. "File:Yin_yang.svg" (graphic file). November 28, 2015. Wikimedia Commons. Retrieved from https://commons.wikimedia.org/wiki/File:Yin_yang.svg.

Allaby, Michael, and Ailisa Allaby, eds. 1999. "Telluric Current." In *A Dictionary of Earth Sciences*. 2nd ed. New York: Oxford University.

Amuecki. 2008. "File:Resonance.png" (graphic file). November 11, 2015. Wikimedia Commons. Retrieved from https://commons.wikimedia.org/wiki/File:Resonance.png.

Association for Asian Studies, Southeast Conference. 1979. *Annals. Volumes 1–5*. Durham, NC: Duke University Press.

Aur, Dorian, and Mandar S. Jog. 2010. *Neuroelectrodynamics: Understanding the Brain Language*. Amsterdam: IOS Press.

Averse. 2007. "File:Diplexer1.jpg" (graphic file). November 13, 2015. Wikimedia Commons. Retrieved from https://en.wikipedia.org/wiki/Microwave#/media/File:Diplexer1.jpg.

Baggott, Jim. 2011. *The Quantum Story: A History in 40 Moments*. New York: Oxford University.

Bailes, K. E. 1990. *Science and Russian Culture in an Age of Revolutions: V.I. Vernadsky and His Scientific School, 1863–1945*. Bloomington: Indiana University.

Bailey, Alice A. 1927. *The Light of the Soul: The Yoga Sutras of Patanjali*. New York: Lucis Publishing.

Bain, George Grantham. 2011. "File:Niels_Bohr_Date_Unverified_LOC.jpg" (graphic file). November 28, 2015. Wikimedia Commons. Retrieved from https://commons.wikimedia.org/wiki/File:Niels_Bohr_Date_Unverified_LOC.jpg.

Becker, Robert O. 1990. *Cross Currents: The Perils of Electropollution and the Promise of Electromedicine*. New York: Jeremy Tarcher.

Becker, Robert O., and Gary Selden. 1985. *The Body Electric: Electromagnetism and the Foundation of Life*. New York: Harper.

Bekenstein, Jacob D. 1973. "Black Holes and Entropy." *Physical Review* 7 (8): 2333–46.

Bell, E. T. 1937. *Men of Mathematics*. New York: Simon and Schuster.

Bell, J. S. 1987. *The Speakable and Unspeakable in Quantum Mechanics*. Cambridge: Cambridge University.

Benenson, Walter, John W. Harris, Horst Stocker, and Lutz Holger, eds. 2006. *Handbook of Physics*. New York: Springer Science.

Berg, Jerome S. 2008. *Broadcasting on the Short Waves, 1945 to Today*. London: McFarland & Company.

Bergson, Henri. 1911. *Creative Evolution*. Translated by Arthur Mitchell. New York: Henry Holt.

Bisson, Terry. 1990. "They're Made Out of Meat." *Omni Magazine* 13 (7): 42–45. New York: General Media.

Blakeslee, Douglas, ed. 1972. *The Radio Amateur's Handbook*. Newington, CT: The American Radio Relay League.

Blinkov, S. M. and I. I. Glezer. 1968. *The Human Brain in Figures and Tables. A Quantitative Handbook*. New York: Plenum.

Block, Ned. 2007. *Consciousness, Function, and Representation: Collected Papers, Vol. 1*. Cambridge: The MIT Press.

Bohm, David. 1951. *Quantum Theory*. New York: Prentiss-Hall.

———. 1952. "A Suggested Interpretation of the Quantum Theory in Terms of 'Hidden Variables,' Vol. 1." *Physical Review* 85 (2): 166–93. Retrieved from http://fma.if.usp.br/~amsilva/Artigos/p166_1.pdf.

———. 1965. *The Special Theory of Relativity*. Philadelphia: John Benjamins.

———. 1978. "The Enfolding-Unfolding Universe: A Conversation with David Bohm." In *The Holographic Paradigm and Other Paradoxes: Exploring the Leading Edge of Science*, edited by Ken Wilber, 44–104. Boulder, CO: Shambhala.

———. 1980. *Wholeness and the Implicate Order*. London: Routledge.

———. 1985. *Unfolding Meaning: A Weekend of Dialogue with David Bohm*. London: Routledge.

———. 1986. "The Implicate Order and the Super-Implicate Order." In *Dialogues with Scientists and Sages: The Search for Unity*, edited by Renée Weber, 23–49. New York: Routledge.

———. 1987a. "Hidden Variables and the Implicate Order." In *Quantum Implications: Essays in Honour of David Bohm,* edited by B. J. Hiley and F. David Peat, 33–45. London: Routledge.

———. 1989. "Meaning and Information." In *The Search for Meaning: The New Spirit in Science and Philosophy*, edited by Paavo Pylkkänen, 43–85. Northamptonshire, England: The Aquarian Press.

———. 1990. *Beyond Limits: A Full Conversation with David Bohm*. Interview by Bill Angelos for Dutch public television. Posted March 5, 2011. Retrieved from http://bohmkrishnamurti.com/beyond-limits/

———. 1990. "A New Theory of the Relationship of Mind and Matter." *Philosophical Psychology* 3 (2): 271–86.

Bohm, David, and Basil J. Hiley. 1993. *The Undivided Universe: An Ontological Interpretation of Quantum Theory*. London: Routledge.

Bohm, David, and J. Krishnamurti. 1985. *The Ending of Time: Where Philosophy and Physics Meet*. New York: Harper Collins.

Bohm, David, and J. Krishnamurti. 1999. *The Limits of Thought: Discussions between J. Krishnamurti and David Bohm*. London: Routledge.

Bohm, David, and F. David Peat. 1987. *Science, Order, and Creativity*. London: Routledge.

Bohm, David, and R. Weber. 1982. "Nature as Creativity." *ReVision* 5 (2): 35–40.

Booth, J. C., S. A. Koren, and Michael A. Persinger. 2005. "Increased Feelings of the Sensed Presence and Increased Geomagnetic Activity at the Time of the Experience During Exposures to Transcerebral Weak Complex Magnetic Fields." *International Journal of Neuroscience* 115 (7): 1039–65.

Born, Irene, trans. 1971. *The Born–Einstein Letters: Friendship, Politics and Physics in Uncertain Times*. New York: Macmillan.

Borsellino, A., and T. Poggio. 1972. "Holographic Aspects of Temporal Memory and Optomotor Responses." *Kybernetik* 10 (1): 58–60.

Brigham, E. O. 2002. *The Fast Fourier Transform*. New York: Prentice-Hall.

Broughton, S. A., and K. Bryan. 2008. *Discrete Fourier Analysis and Wavelets: Applications to Signal and Image Processing*. New York: Wiley.

Browder, Andrew. 1996. *Mathematical Analysis: An Introduction*. New York: Springer-Verlag.

Bruskiewich, Patrick. 2014. *Max Planck and Black-Body Radiation*. Vancouver: Pythagoras Publishing.

Carr1, Bernard J. and Steven B. Giddings. 2005. "Quantum Black Holes." *Scientific American* 292 (5): 30–35.

Cazenave, Michel, ed. 1984. *Science and Consciousness: Two Views of the Universe, Edited Proceedings of the France-Culture and Radio-France Colloquium, Cordoba, Spain*. Oxford: Pergamon.

Chalmers, David J. 1995. "Facing Up to the Problem of Consciousness." *Journal of Consciousness Studies* 2 (3): 200–19.

———. 2010. *The Character of Consciousness*. New York: Oxford University.

Chaudhuri, Haridas. 1954. *The Philosophy of Integralism: The Metaphysical Synthesis in Sri Aurobindo's Teaching*. Pondicherry, India: Sri Aurobindo Ashram Trust.

———. 1960. “The Integral Philosophy of Sri Aurobindo.” In *The Integral Philosophy of Sri Aurobindo: A Commemorative Symposium*, edited by Haridas Chaudhuri and Frederic Spiegelberg, 17–34. London: George Allen & Unwin.

———. 1969. *The Integral Philosophy of Sri Aurobindo*. London: George Allen & Unwin.

Chen, Frances F. 2006. *Introduction to Plasma Physics and Controlled Fusion: Vol 1. Plasma Physics*. 2nd ed. New York: Springer.

Cheney, Margaret. 1981. *Tesla: Man Out of Time*. New York: Dorset Press.

Chomsky, Noam. 2000. *New Horizons in the Study of Language and Mind*. Cambridge, England: Cambridge University.

Clark, Walter. 1939. *Photography by Infrared—Its Principles and Applications*. New York: John Wiley & Sons.

Cohn, Stephen M., 2007. “Near-Infrared Spectroscopy: Potential Clinical Benefits in Surgery.” *Journal of the American College of Surgeons* 205 (2): 322–32. doi: http://dx.doi.org/10.1016/j.jamcollsurg.2007.02.024.

Collister, Rupert Clive. 2010. *A Journey in Search of Wholeness and Meaning*. New York: Peter Lang.

Cook, David M. 2002. *The Theory of the Electromagnetic Field*. Englewood Cliffs, NJ: Prentice-Hall.

Crease, Robert P. 2008. *The Great Equations: Breakthroughs in Science from Pythagoras to Heisenberg*. New York: W. W. Norton.

Darekk2. 2012. "File:Water_infrared_absorption_coefficient_large.gif" (graphic file). November 28, 2015. Wikimedia Commons. Retrieved from https://en.wikipedia.org/wiki/File:Water_infrared_absorption_coefficient_large.gif.

Darling, David J. 2004. *The Universal Book of Mathematics: From Abracadabra to Zeno's Paradoxes*. Hoboken, NJ: Wiley.

Dawson, Lorne L. 2006. *Comprehending Cults: The Sociology of New Religious Movements*. New York: Oxford University.

Deacon, Terrence W. 2010. "What Is Missing from Theories of Information." In *Information and the Nature of Reality: From Physics to Metaphysics*, edited by Paul Davies and Niels Henrik Gregersen, 123–42. Cambridge, England: Cambridge University.

———. 2012. *Incomplete Nature: How Mind Emerged from Matter*. New York: W. W. Norton.

De Valois, Karen K., and Russell L. De Valois. 1988. *Spatial Vision*. New York: Oxford University.

Dennett, Daniel C. 1991. *Consciousness Explained*. New York: Back Bay Books.

Dewey, B. 1985. *The Theory of Laminated Spacetime*. Inverness, CA: Bartholomew.

Dewynne, Dustin. 2012. “File:Dualism-vs-Monism.png” (graphic file). November 27, 2015. Wikimedia Commons. Retrieved from https://commons.wikimedia.org/wiki/File:Dualism-vs-Monism.png.

Differenxe. 2010. “File:Sawtooth_Fourier_Analysys.svg” (graphic file). October 22, 2015. Wikimedia Commons. Retrieved from http://commons.wikimedia.org/wiki/File:Sawtooth_Fourier_Analysys.svg.

Dorf, Richard C., ed. 1997. *The Electrical Engineering Handbook*. 2nd ed. Boca Raton, FL: CRC Press.

Drachman, David A. 2005. “Do We Have Brain to Spare?” *Neurology* 64 (6): 2004–5. doi: http://dx.doi.org/10.1212/01.WNL.0000166914.38327.BB.

Dunwell, Frances F. 1980. *The Hudson: America’s River*. New York: Columbia University.

Edelman, Gerald M., and Giulio Tononi. 2000. *A Universe of Consciousness: How Matter Becomes Imagination*. New York: Basic Books.

Edmondson, Amy C. 1987. *A Fuller Explanation: The Synergetic Geometry of R. Buckminster Fuller*. Boston: Birkhäuser.

Einstein, Albert. 1979. *Autobiographical Notes*. Peru, IL: Carus.

Eliot, Thomas Stearns. 1943. *Four Quartets*. New York: Harcourt Brace.

———. 1976. *Collected Poems 1909–1962*. New York: Harcourt Brace.

Fechner, Gustav Theodor. 1946. *Religion of a Scientist: Selections from Gustav Theodor Fechner*, New York: Pantheon.

Fellmann, E. A. 2007. *Leonhard Euler*. Translated by E. Gautschi. Basel, Germany: Birkhauser.

Ferrer, Jorge, and Jacob Sherman, eds. 2008. *The Participatory Turn: Spirituality, Mysticism, Religious Studies*. Albany: State University of New York.

Feuerstein, Georg. 1987. *Structures of Consciousness: The Genius of Jean Gebser—An Introduction and Critique*. Lower Lake, CA: Integral.

Feynman, Richard, Robert Leighton, and Matthew Sands. 1964. *The Feynman Lectures on Physics, Vol. 1*. Reading, MA: Addison-Wesley.

Fields, R. Douglas. 2009. *The Other Brain*. New York: Simon & Schuster.

Flanagan, Owen. 1997. "Conscious Inessentialism and the Epiphenomenalist Suspicion." In *The Nature of Consciousness: Philosophical Debates*, edited by Ned Block, Owen Flanagan, and Güven Güzeldere, 357–73. Cambridge, MA: MIT Press.

Fourier, Jean Baptiste Joseph. 1822. *The Analytic Theory of Heat*. Paris: Firmin Didot Père et Fils.

Fulvio314. 2013. "File:Coat_of_Arms_of_Niels_Bohr.svg" (graphic file). November 28, 2015. Wikimedia Commons. Retrieved from https://commons.wikimedia.org/wiki/File:Coat_of_Arms_of_Niels_Bohr.svg.

Gabor, Dennis. 1946. "Theory of Communication." *Journal of the Institute of Electrical Engineers* 93: 429–41.

Gao, Shan. 2014. *Dark Energy: From Einstein's Biggest Blunder to the Holographic Universe.* 2nd ed. Seattle, WA: Amazon Kindle Direct.

Garay, Luis J. 1995. "Quantum Gravity and Minimum Length." *International Journal of Modern Physics* 10 (2): 145–65.

Gebser, Jean. 1949. *The Ever-Present Origin: Part One: Foundations of the Aperspectival World.* Translated by J. Keckeis. Stuttgart, Germany: Deutsche Verlags-Anstalt.

———. 1956 (1996). "Cultural Philosophy as Method and Venture." Translated by Georg Feuerstein. *Integrative Explorations: Journal of Culture and Consciousness* 3 (1): 77–82. Retrieved from http://static1.squarespace.com/static/535ef5d8e4b0ab57db4a06c7/t/541f74a0e4b0394ddbf5a040/1411347616917/integrative_explorations_3.pdf.

Geiger, Rudolf, Robert H. Aron, and Paul Todhunter. 2003. *The Climate Near the Ground.* Lanham, MD: Rowman and Littlefield.

Gidley, Jennifer. 2007. "The Evolution of Consciousness as a Planetary Imperative: An Integration of Integral Views." *Integral Review* 3 (5): 4–226. Retrieved from http://integral-review.org/pdf-template-issue.php?pdfName=issue_5_gidley_the_evolution_of_consciousness_as_a_planetary_imperative.pdf.

Glatzmaier. 2007. "File:Geodynamo_Between_Reversals.gif" (graphic file). November 12, 2015. Wikimedia Commons. Retrieved from https://commons.wikimedia.org/wiki/File:Geodynamo_Between_Reversals.gif.

Globus, Gordon G. 2006. *The Transparent Becoming of World: A Crossing Between Process Philosophy and Quantum Neurophilosophy*. Philadelphia: John Benjamins.

Goswami, Amit. 2000. *The Visionary Window: A Quantum Physicist's Guide to Enlightenment*. Wheaton, IL: Quest Books.

Gott, J. Richard III, Mario Jurić, David Schlegel, and Fiona Hoyle. 2005. "A Map of the Universe." *The Astrophysics Journal* 624 (2): 463–514.

Gray, Henry. 1918. "File:Gray754.png" (graphic file). October 31, 2015. Wikimedia Commons. Retrieved from https://commons.wikimedia.org/wiki/File:Gray754.png.

Haas, Jonathan. 2012. "File:Neurons_uni_bi_multi_pseudouni.svg" (graphic file). October 31, 2015. Wikimedia Commons. Retrieved from http://commons.wikimedia.org/wiki/File:Neurons_uni_bi_multi_pseudouni.svg.

Haisch, Bernard. 2010. *The Purpose-Guided Universe: Believing in Einstein, Darwin, and God*. Pompton Plains, NJ: New Page Books.

Hameroff, Stuart R. 2015. "Is Your Brain Really a Computer, or Is It a Quantum Orchestra?" *Huffington Post*, July 9. Retrieved from http://www.huffingtonpost.com/stuart-hameroff/is-your-brain-really-a-co_b_7756700.html.

Hameroff, Stuart, and Roger Penrose. 1996. "Conscious Events as Orchestrated Space-Time Selections." *Journal of Consciousness Studies* 3 (1): 35–53.

Hameroff, Stuart R., Travis J. A. Craddock, and Jack A. Tuszynski. 2014. "Quantum Effects in the Understanding of Consciousness." *Journal of Integrative Neuroscience* (13) 2: 229–52. doi:10.1142/S0219635214400093.

Harnad, Stevan. 1994. "Why and How We Are Not Zombies." *Journal of Consciousness Studies* 1 (1): 18–23.

Havelka, D., M. Cifra, O. Kucera, J. Pokorny, and J. Vrba. 2011. "High-frequency Electric Field and Radiation Characteristics of Cellular Microtubule Network." *Journal of Theoretical Biology* 286 (7): 31–40.

Heehs, Peter. 2008. *The Lives of Sri Aurobindo*. New York: Columbia University.

Herculano-Houzel, Suzana. 2009. "The Human Brain in Numbers: A Linearly Scaled-up Primate Brain." Frontiers in Human Neuroscience (3) 31: 1–11. doi: 10.3389/neuro.09.031.2009

Hertz, Heinrich. 1893. *Electric Waves: Being Researches On the Propagation of Electric Action with Finite Velocity Through Space*. London: MacMillan.

Hiley, B. J., and F. David Peat, eds. 1987. *Quantum Implications: Essays in Honour of David Bohm*. London: Routledge.

Hiley, B. J., and F. David Peat, eds. 1987. "The Development of David Bohm's Ideas from the Plasma to the Implicate Order." In *Quantum Implications: Essays Honour of David Bohm*, edited by B. J. Hiley and F. David Peat, 1–32. London: Routledge.

Horne, Alistair. 1962. *The Price of Glory: Verdun 1916*. New York: St. Martin's Press.

Howell, Kenneth B. 2001. *Principles of Fourier Analysis*. Boca Raton, FL: Chapman & Hall.

Ishikawa, Hiroaki, and Wallace F. Marshall. "Ciliogenesis: Building the Cell's Antenna." *Nature Reviews Molecular Biology* 12 (4): 222-34. dos:10.1038/nrm3085

Jahoe. 2012. "File:Electromagnetic-Spectrum.png" (graphic file). November 28, 2015. Wikimedia Commons. https://commons.wikimedia.org/wiki/File:Electromagnetic-Spectrum.png.

Jbarta. 2014. "File:Max_Planck_1878.GIF" (graphic file). November 28, 2015. Wikimedia Commons. Retrieved from https://commons.wikimedia.org/wiki/File:Max_Planck_1878.GIF.

Jibu, Mari, and Kunio Yasue. 1995. *Quantum Brain Dynamics and Consciousness*. Philadelphia: John Benjamins.

———. 2003. "Quantum Brain Dynamics and Quantum Field Theory." In *Brain and Being: At the Boundary Between Science, Philosophy, Language and Arts*, edited by Gordon Globus, Karl Pribram, and Giuseppe Vitiello, 267–90. Philadelphia: John Benjamins.

Johnston, Sean F. 2006. *Holographic Visions: A History of New Science*. New York: Oxford University.

Jung, C. G. 1968. *Psychology and Alchemy*. Vol. 12 of *The Collected Works of C. G. Jung*. Edited and translated by Gerald Adler and R. F. C. Hull. 2nd ed. Princeton, NJ: Princeton University.

———. (1946) 1969. "On the Nature of the Psyche." In Vol. 8 of *The Collected Works of C. G. Jung*, translated by R. F. C. Hull, 159–234. 2nd ed. Princeton, NJ: Princeton University.

Kachris, Christoforos, Keren Bergman, and Ioannis Tomkos. 2012. *Optical Interconnects for Future Data Center Networks*. New York. Springer.

Kafatos, Menas, Rudolph E. Tanzi, and Deepak Chopra. 2011. "How Consciousness Becomes the Physical Universe." *Journal of Cosmology* (14): 1318–1328. Retrieved from http://journalofcosmology.com/Consciousness140.html.

Köhler, Wolfgang. 1940. *Dynamics in Psychology: Vital Applications of Gestalt Psychology*. New York: W. W. Norton.

———. 1969. *The Task of Gestalt Psychology*. New Jersey: Princeton University.

Köhler, Wolfgang, and Mary Henle. 1971. *The Selected Papers of Wolfgang Köhler*. New York: W. W. Norton.

Kuehn, Kerry. 2014. *A Student's Guide Through the Great Physics Texts, Vol. 1: The Heavens and The Earth*. New York: Springer.

Kuo, Franklin. 1962. *Network Analysis and Synthesis*. New York: John Wiley & Sons.

Lashley, Karl. 1951. "An Examination of the Electric Field Theory of Cerebral Integration." *Psychological Review* 58: 123–36.

———. 1950. "In Search of the Engram." *Symposium of the Society for Experimental Biology* 4: 454–82.

László, Ervin. 2006. *Science and the Re-Enchantment of the Cosmos: The Rise of the Integral Vision of Reality*. Rochester, VT: Inner Traditions.

———. 2007. *Science and the Akashic Field: An Integral Theory of Everything*. Rochester, VT: Inner Traditions.

———. 2014a. *The Immortal Mind: Science and the Continuity of Consciousness beyond the Brain*. Rochester, VT: Inner Traditions.

———. 2014b. *The Self-Actualizing Cosmos: The Akasha Revolution in Science and Human Consciousness*. Rochester, VT: Inner Traditions.

Leckie, Robert. 1987. *Delivered from Evil: The Saga of World War II*. New York: Harper & Row.

Leith, Emmet N., and J. Upatnieks. 1962. "Reconstructed Wavefronts and Communication Theory." *Journal of the Optical Society of America* 52 (10): 1123–30.

———. 1965. "Photography by Laser." *Scientific American* 212 (6): 24–35.

Lilly, John. 1967. *The Mind of the Dolphin: A Nonhuman Intelligence.* New York: Doubleday.

———. 1977. *The Deep Self: Consciousness Exploration in the Isolation Tank.* New York: Simon & Schuster.

Livio, Mario. 2003. *The Golden Ratio: The Story of Phi, the World's Most Astonishing Number.* New York: Broadway Books.

Lundqvist, Stig, ed. 1992. *Nobel Lectures, Physics 1971–1980.* Singapore: World Scientific.

Lyon, Richard F. "File:Sine_wavelength.svg" (graphic file). November 22, 2015. Wikimedia Commons. Retrieved from https://commons.wikimedia.org/wiki/File:Sine_wavelength.svg.

MacKenna, Stephen. 1992. *Plotinus: The Enneads.* New York: Larson Publications.

Malinski, Tadeusz. 1960. *Chemistry of the Heart.* Ohio: Biochemistry Research Laboratory. Retrieved from http://hypertextbook.com/facts/2003/IradaMuslumova.shtml

Mandelbrot, Benoit. 1986. "Fractals and the Rebirth of Iteration Theory." In *The Beauty of Fractals: Images of Complex Dynamical Systems*, by H.-O. Peitgen and P. H. Richter, 151–60. Berlin: Springer-Verlag.

McCraty, Rollin. 2003. *The Energetic Heart: Bioelectromagnetic Interactions Within and Between People*. Boulder Creek, CA: Institute of HeartMath.

McCraty, Rollin, M. Atkinson, D. Tomasino, and R. T. Bradley. 2009. "The Coherent Heart: Heart–Brain Interactions, Psychophysiological Coherence, and the Emergence of System-Wide Order." *Integral Review* 5 (9): 10–115.

McCraty, Rollin, Annette Deyhle, and Doc Childre. 2012. "The Global Coherence Initiative: Creating a Coherent Planetary Standing Wave." *Global Advances in Health and Medicine* 1 (1): 64–77. Retrieved from https://www.heartmath.org/assets/uploads/2015/01/gci-creating-a-coherent-planetary-standing-wave.pdf.

McFadden, Johnjoe. 2002a. "The Conscious Electromagnetic Information (CEMI) Field Theory: The Hard Problem Made Easy." *Journal of Consciousness Studies* 9 (8): 45–60.

———. 2002b. "Synchronous Firing and Its Influence on the Brain's Electromagnetic Field: Evidence for an Electromagnetic Field Theory of Consciousness." *Journal of Consciousness Studies* 9 (4): 23–50.

———. 2006. "The CEMI Field Theory." In *The Emerging Physics of Consciousness*, edited by Jack A. Tuszynski, 385–404. Berlin: Springer-Verlag.

———. 2007. "Conscious Electromagnetic Field Theory." *NeuroQuantology* 5 (3): 262–70.

McGinn, Colin. 1997. "Consciousness and Space." In *Explaining Consciousness: The Hard Problem*, edited by Jonathan Shear, 97–108. Boston: MIT.

McIntosh, Steve. 2007. *Integral Consciousness and the Future of Evolution*. St. Paul, MN: Paragon House.

Merrell-Wolff, Franklin. 1973. *The Philosophy of Consciousness Without an Object: Reflections on the Nature of Transcendental Consciousness*. New York: Julian Press.

Milonni, Peter W., and Joseph E. Eberly. 2010. *Laser Physics*. Hoboken, NJ: John Wiley & Sons.

Morgan, Conway Lloyd. 1978. *Emergent Evolution: Gifford Lectures, 1921–22*. New York: Simon & Schuster.

Morin, Edgar. 1999. *Seven Complex Lessons in Education for the Future*. Translated by Nidra Poller. Paris: UNESCO.

Mysid. 2008. "File:Atmospheric_electromagnetic_opacity.svg" (graphic file). November 28, 2015. Wikimedia Commons. https://commons.wikimedia.org/wiki/File:Atmospheric_electromagnetic_opacity.svg.

National Multiple Sclerosis Society. n.d. "Cerebrospinal Fluid (CSF)." Retrieved from http://www.nationalmssociety.org/Symptoms-Diagnosis/Diagnosing-Tools/Cerebrospinal-Fluid-(CSF).

Netter, F. H. 1972. *A Compilation of Paintings of the Normal and Pathologic Anatomy of the Nervous System*. Summit, NJ: CIBA.

Neville, Katherine. 1992. "Saral and David Bohm, Prague, June 1992" (photograph). Retrieved from http://www.karlpribram.com/photos/.

Nilson, Arthur R., and J. L. Hornung. 1943. *Practical Radio Communication: Principles, Systems, Equipment, Operation, Including Very High and Ultra High Frequencies and Frequency Modulation*. 2nd ed. New York: McGraw-Hill.

Nishikawa, K., and M. Wakatani. 2000. *Plasma Physics*. Berlin: Springer-Verlag.

Nunez, Paul L. 2010. *Brain, Mind, and the Structure of Reality*. New York: Oxford University.

Oates, B. 1971. *Celebrating the Dawn: Maharishi Mahesh Yogi and the TM Technique*. New York: Putnam Books.

Ouspensky, P. D. 1949. *In Search of the Miraculous: Fragments of an Unknown Teaching*. London: Harcourt.

Oyster, Clyde W. 1999. *The Human Eye: Structure and Function*. Sunderland, MA: Sinauer Associates.

Peat, F. David. 1997. *Infinite Potential: The Life and Times of David Bohm*. Reading, MA: Addison-Wesley.

Pelekanos. (1478) 2007. "File:Serpiente_alquimica.jpg" (graphic file). File uploaded April 10 by Carlos Adanero. Wikimedia Commons. Retrieved from https://commons.wikimedia.org/wiki/File:Serpiente_alquimica.jpg.

Penfield, Wilder. 1975. *The Mystery of the Mind: A Critical Study of Consciousness and the Human Brain*. Princeton, NH: Princeton University.

Penrose, Sir Roger. 1989. *The Emperor's New Mind: Concerning Computers, Minds and the Laws of Physics*. New York: Oxford University.

Penrose, R., Stuart Hameroff, and S. Kak, eds. 2011. *Consciousness and the Universe: Quantum Physics, Evolution, Brain & Mind* (Contents selected from Volumes 3 and 14, *Journal of Cosmology*). Cambridge: Cosmology Science.

Persinger, Michael A. 2014. "Schumann Resonance Frequencies Found Within Quantitative Electroencephalographic Activity: Implications for Earth-Brain Interactions." *International Letters of Chemistry, Physics and Astronomy* 30: 24–32. doi:10.18052/www.scipress.com/ILCPA.30.24.

Pizzi, Rita, Giuliano Strini, Silvia Fiorentini, Valeria Pappalardo, and Massimo Pregnolato. 2010. "Evidences of New Biophysical Properties of Microtubules." In *Artificial Neural Networks*, edited by Seoyun J. Kwon. New York: Nova Science Publishers, 1–17. Retrieved from https://air.unimi.it/retrieve/handle/2434/167480/168890/evidences.pdf.

Planck, Max. 1901. "On the Law of Distribution of Energy in the Normal Spectrum." *Annalen der Physik*. 309 (3): 553–63.

Pockett, Susan. 2000. *The Nature of Consciousness: A Hypothesis*. Lincoln, NE: Writers Club.

———. 2007. "Difficulties with the Electromagnetic Field Theory of Consciousness: An Update." *NeuroQuantology* 5 (3): 271–75. Retrieved from http://www.neuroquantology.com/index.php/journal/article/view/136/136 (login required).

Pribram, Karl H. 1962. "The Neuropsychology of Sigmund Freud." In *Experimental Foundations of Clinical Psychology*, edited by Arthur J. Bachrach, 442–68. New York: Basic Books.

———. 1971. *Languages of the Brain: Experimental Paradoxes and Principles in Neuropsychology*. Englewood Cliffs, NJ: Prentice-Hall.

———. 1982. "What the Fuss is All About." In *The Holographic Paradigm and Other Paradoxes*, edited by Ken Wilber, 27–34. Boulder: Shambhala.

———. 1984. "Mind, Brain and Consciousness: The Organization of Competence and Conduct." In *Science and Consciousness: Two Views of the Universe*, edited by Julian Davidson and Richard Davidson, 115–32. New York: Springer.

———. 1990. "Prolegomenon for a Holonomic Brain Theory." In *Synergetics of Cognition*, edited by H. Haken, 150–84. Berlin: Springer-Verlag.

———. 1991. *Brain and Perception: Holonomy and Structure in Figural Processing*. Hillsdale, NJ: Lawrence Erlbaum.

———. 1994. "What Is Mind that the Brain May Order It?" In *Proceedings of Symposia in Applied Mathematics: Proceedings of the Norbert Wiener Centenary Congress, Vol. 52*, edited by V. Mandrekar and P. R. Masani, 301–29. Providence, RI: American Mathematical Society.

———. 2004a. "Brain and Mathematics." In *Brain and Being: At the Boundary Between Science, Philosophy, Language and Arts*, edited by Gordon Globus, Karl Pribram, and Giuseppe Vitiello, 215–40. Philadelphia: John Benjamins.

———. 2004b. "Consciousness Reassessed." *Mind and Matter* 2 (1): 7–35.

———. 2011. "Karl Pribram: Bibliography." Retrieved from http://www.karlpribram.com/bibliography/.

———. 2013. *The Form Within: My Point of View*. Westport, CT: Prospecta Press.

Pribram, Karl H., A. Sharafat, and G. Beekman. 1984. "Frequency Encoding in Motor Systems." In *Human Motor Actions: Bernstein Reassessed*, edited by H. T. A. Whiting, 121–56. Amsterdam: Elsevier Science Publishers.

Prigogine, Ilya. 2015. *Modern Thermodynamics: From Heat Engines to Dissipative Structures*. Hoboken, NJ: John Wiley & Sons.

Pylkkänen, Paavo. 2007. *Mind, Matter, and the Implicate Order*. New York: Springer.

Radhakrishnan, Sarvepalli, ed. 1952. *History of Philosophy Eastern and Western, Vol. 2.* London: George Allen & Unwin.

Ramon y Cajal, Santiago. 2007. "File: Purkinje_cell_by_Cajal.png" (graphic file). January 30, 2016. Wikimedia Commons. Retrieved from https://commons.wikimedia.org/wiki/File:Purkinje_cell_by_Cajal.png.

Rendek, Kimberly N., Raimund Fromme, Ingo Grotjohann, and Petra Fromme. 2013. "Self-Assembled Three-Dimensional DNA Nanostructure." *Acta Crystallographica Section F: Structural Biology and Crystallization Communications* 69 (2): 141–46. doi:10.1107/S1744309112052128.

Rescher, Nicholas, ed. 1991. *G. W. Leibniz's Monadology: An Edition for Students.* Pittsburgh: University of Pittsburgh.

Ringbauer, M., B. Duffus, C. Branciard, E. G. Cavalcanti, A. G. White, and A. Fedrizzi. 2015. "Measurements On the Reality of the Wavefunction." *Nature Physics* 11 (2): 249–254. doi:10.1038/nphys3233.

Romanes, G. J., ed. 1964. *Cunningham's Textbook of Anatomy.* 10th ed. New York: Oxford University.

Romary. 2006. "File:Come pariou.jpg" (graphic file). January 29, 2016. Wikimedia Commons. Retrieved from https://commons.wikimedia.org/wiki/File:Come_pariou.jpg.

Rose, D. 2006. *Consciousness: Philosophical, Psychological and Neural Theories*. New York: Oxford University.

Ruhenstroth-Bauer, Gerhard. 1993. "Influence of the Earth's Magnetic Field on Resting and Activated EEG Mapping in Normal Subjects." *International Journal of Neuroscience* 73 (3): 331–49.

Runehov, Anne, and Luis Oviedo, eds. 2013. *Encyclopedia of Sciences and Religions*. Dordrecht, Netherlands: Springer Netherlands.

Ruppert, L. 1956. *History of the International Electrotechnical Commission*. Geneva: Central Bureau of the IEC.

Samson, Paul R., and David Pitt, eds. 1999. *The Biosphere and Noosphere Reader: Global Environment, Society and Change*. New York: Routledge.

Schrock, Karen. 2007. "An Opera Singer's Piercing Voice Can Shatter Glass." *Scientific American*, August 23. Retrieved from http://www.scientificamerican.com/article/fact-or-fiction-opera-singer-can-shatter-glass/

Sentman, Davis D. 1995. "Schumann Resonances." In *Handbook of Atmospheric Electrodynamics, Vol. 1*, edited by Hans Volland, 267–96. Boca Raton, FL: CRC Press.

Shannon, C. E. 1948. "A Mathematical Theory of Communication." *Bell System Technical Journal* 27: 623–56.

Sheldrake, Rupert. 1981. *A New Science of Life: The Hypothesis of Morphic Resonance*. Rochester, VT: Park Street.

———. 1988. *The Presence of the Past: Morphic Resonance and the Habits of Nature*. New York: Times Books.

———. 1989. *Morphic Resonance: The Nature of Formative Causation*. Rochester, VT: Park Street.

Skrbina, David. 2007. *Panpsychism in the West*. Cambridge, MA: MIT Press.

Smolin, Lee. 2013. *Time Reborn*. London: Penguin Books.

Sperry, Roger W., N. Miner, and R.E. Myers. 1955. "Visual Pattern Perception Following Subpial Slicing and Tantalum Wire Implantations in the Visual Cortex." *Journal of Comparative and Physiological Psychology* 48 (1): 50–58.

Squire, Larry R., ed. 1998. *The History of Neuroscience in Autobiography, Vol. 2*. London: Academic Press.

Stanford Daily. 1977. "Scientists to Speak." 172 (2): 8.

Stapp, Henry P. 2009. *Mind, Matter, and Quantum Mechanics*. 3rd ed. New York: Springer-Verlag.

Steiner, Rudolf. 1959. *Cosmic Memory: Prehistory of Earth and Man*, translated by K. E. Zimmer. San Francisco: Harper & Row.

Stuart, C.I.J., Y. Takahashi, and H. Umezawa .1979. "Mixed System Brain Dynamics: Neural

Memory as a Macroscopic Ordered State." *Foundations of Physics* (9): 301–7.

Susskind, Leonard. 2008. *The Black Hole War: My Battle with Stephen Hawking to Make the World Safe for Quantum Mechanics*. New York: Little, Brown and Company.

Svdmolen. 2005. "File:Red_White_Blood_cells.jpg" (graphic file). November 12, 2015. Wikimedia Commons. Retrieved from https://en.wikipedia.org/wiki/File:Red_White_Blood_cells.jpg.

Taimni, I. K. 1969. *Man, God and the Universe*. Madras, India: The Theosophical Society.

Talbot, Michael. 1992. *The Holographic Universe*. New York: Harper Collins.

Telsa, Faraday. 2014. "File:Assembly_of_microscale_beads_on_Faraday_waves.gif" (graphic file). February 7, 2016. Wikimedia Commons. https://upload.wikimedia.org/wikipedia/commons/4/40/Assembly_of_microscale_beads_on_Faraday_waves.gif.

Tjlaxs. 2005. "File:YoungJamesClerkMaxwell.jpg" (graphic file). November 28, 2015. Wikimedia Commons. https://commons.wikimedia.org/wiki/File:YoungJamesClerkMaxwell.jpg.

Todeschi, Kevin J. 1998. *Edgar Cayce on the Akashic Records: The Book of Life*. Virginia Beach, VA: A. R. E. Press.

Turgeon, Mary Louise. 2004. *Clinical Hematology: Theory and Procedures*. Baltimore, MD: Lippincott Williams & Wilkins.

Tudzynski, P., T. Correia, and U. Keller. 2001. "Biotechnology and Genetics of Ergot Alkaloids." *Applied Microbiology and Biotechnology*, 57: 593–605. doi:10.1007/s002530100801.

Tuszyński, Jack A., ed. 2006. *The Emerging Physics of Consciousness*. New York: Springer-Verlag.

Van Dokkum, Pieter G., and Charlie Conroy. 2010. "A Substantial Population of Low-Mass Stars in Luminous Elliptical Galaxies." *Nature* 468 (7326): 940–42. doi:10.1038/nature09578.

Vejvoda, Stanislav. 2003. *17th International Conference on Structural Mechanics in Reactor Technology*. Prague, Czech Republic: Czech Standard Institute.

Vernadsky, Vladimir Ivanovich. 1998. *The Biosphere*. Translated by D. B. Langmuir. New York: Springer-Verlag.

Walker, J. Samuel. 2004. *Three Mile Island: A Nuclear Crisis in Historical Perspective*. Berkeley: University of California Press.

Warren, Stephen G., and Richard E. Brandt. 2008. "Optical Constants of Ice from the Ultraviolet to the Microwave: A Revised Compilation." *Journal of Geophysical Research*, 113 (D14): 1047–57.

Wasson, Tyler, and Gert Brieger. 1987. *Nobel Prize Winners: A Biographical Dictionary*. New York: H. W. Wilson.

Weber, Renée. 1982. "The Physicist and the Mystic—Is a Dialogue Between Them Possible?" In *The Holographic Paradigm and Other Paradoxes: Exploring the Leading Edge of Science*, edited by Ken Wilber, 187–214. Boulder, CO: Shambhala.

Weisenberger, Drew. n.d. "Jefferson Lab Questions and Answers: How Many Atoms Are There in the World?" Retrieved January 20, 2016 from http://education.jlab.org/qa/mathatom_05.html

Wheeler, John Archibald. 1990. *Information, Physics, Quantum: The Search for Links*. Austin, TX: University of Texas.

———. 1990. *A Journey into Gravity and Spacetime*. New York: Scientific American.

———. 1998. *Geons, Black Holes, and Quantum Foam: A Life in Physics*. New York: W. W. Norton.

Whicher, Ian. 1997. "Nirodha, Yoga Praxis, and the Transformation of the Mind." *Journal of Indian Philosophy* 25 (1): 1–67.

Whitehead, Alfred North. 1978. *Process and Reality: An Essay in Cosmology. Gifford Lectures, 1927–28*. New York: Simon & Schuster.

Wick, Manfred, Wulf Pinggera, and Paul Lehmann. 2003. *Clinical Aspects and Laboratory Iron Metabolism*. Vienna, Austria: Springer-Verlag.

Wiener, Norbert. 1948. *Cybernetics or Control and Communication in the Animal and the Machine*. Cambridge, MA: MIT Press.

Wilber, Ken, ed. 1982. *The Holographic Paradigm and Other Paradoxes: Exploring the Leading Edge of Science*. Boulder, CO: Shambhala.

———, ed. 1985. *Quantum Questions: Mystical Writings of the World's Great Physicists*. Boston: Shambhala.

———. 1996. *A Brief History of Everything*. Boston: Shambhala.

———. 1997. *Integral Spirituality: A Startling New Role for Religion in the Modern and Postmodern World*. Boston: Integral Books.

Wiltschko, Wolfgang, and Roswitha Wiltschko. 2008. "Magnetic Orientation and Magnetoreception in Birds and Other Animals." *Journal of Comparative Physiology A: Neuroethology, Sensory, Neural, and Behavioral Physiology* 191 (8): 675–93. doi:10.1007/s00359-005-0627-7.

Wiseman, Howard M., Fuwa Maria, Shuntaro Takeda, Marcin Zwierz, and Akira Furusawa. 2015. "Experimental Proof of Nonlocal Wavefunction Collapse for A Single Particle Using Homodyne Measurements." *Nature Communications* 6: 192–203. Retrieved from http://www.nature.com/ncomms/2015/150324/ncomms7665/full/ncomms7665.html.

Wong, Eva. 1997. *The Shambhala Guide to Taoism*. Boston: Shambhala.

Woolf, N. J. 2006. "Microtubules in the Cerebral Cortex: Role in Memory and Consciousness." In *The Emerging Physics of Consciousness*, edited by Jack A. Tuszyński, 49–94. New York: Springer-Verlag.

Yau, Shing-Tung, and Steve Nadis. 2010. *The Shape of Inner Space: String Theory and the Geometry of the Universe's Hidden Dimensions*. New York: Basic Books.

Zimmer C. 2009. "Origins: On The Origin of Eukaryotes." *Science* 325 (5941): 666–68.

Zizzi, Paola. 2006. "Consciousness and Logic in a Quantum Computing Universe." In *The Emerging Physics of Consciousness*, edited by Jack A. Tuszyński, 457–81. New York: Springer-Verlag.

APPENDIX:
The Holofield Theory of Ervin László

The holofield theory of Ervin László correlates well with an essential concept of the holoflux theory of consciousness developed in this book, specifically in the postulated existence of two domains or dimensions of reality, an "A-dimension" equivalent to Bohm's implicate order or Pribram's frequency flux domain, and an "M-dimension," equivalent to Bohm's explicate, space–time material domain.[343] A graduate of the Sorbonne and a professor of Philosophy, Systems Sciences, and Future Studies, László had been twice nominated for the Nobel Peace Prize; he writes of the universe, including human beings, appearing to be "nonlocally" coherent, part of a unified field of consciousness.[344]

In agreement with John Archibald Wheeler, who held that information is even more fundamental to the universe than energy, László describes an "information field," which he also calls the "A-field," lying outside of space–time and yet interacting dynamically with material manifesting in space and time.[345] In reference to this A-field, or A-dimension, László frequently uses the term *Akasha*, referencing the Sanskrit word (Ākāśa), described in Vedic texts in India as early as 5,000 BCE, in which Akasha is closely associated with

---

[343] László, *Science and the Re-Enchantment of the Cosmos*, 34–35.

[344] László, *The Self-Actualizing Cosmos*, 44–45.

[345] László, *An Integral Theory of Everything*, 67.

vibration.[346] In the Vedas its function is identified with *shabda,* the first vibration, the first ripple beginning our universe, and also with *spanda,* which has been described as "vibration/movement of consciousness."[347] The Akasha has been described by an Indian professor of inorganic chemistry, I.K. Taimni, as a sort of space out of which integrated energy vibrations emanate (much as holoflux emanates from out of an implicate domain into space–time as electromagnetic waves):

> There is . . . a mysterious integrated state of vibration from which all possible kinds of vibrations can be derived by a process of differentiation. That is called *Nāda* in Sanskrit. It is a vibration in a medium called *Ākāśa* which may be translated as "space" in English. But the conception of *Ākāśa* . . . is quite different from that of Science. It is not mere empty space but space which, though apparently empty, contains within itself an infinite amount of potential energy . . . . This infinite potentiality for producing vibrations of different kinds in any intensity or amount is due to the fact that at the back of *Ākāśa* or hidden within it, is consciousness.[348]

In László cosmology, the universe is the interaction of two dimensions, an "A-dimension" and an "M-dimension" ("M" for "material"). The M-dimension contains that which manifests (i.e., the

[346] Whicher, "Nirodha, Yoga Praxis, and the Transformation of the Mind," 67.

[347] Ibid.

[348] Taimni, *Man, God and the Universe,* 203.

cosmos, the galaxies in space–time), and is that upon which material science has been primarily focused. The M-dimension can be seen as congruent with David Bohm's explicate order. László's other dimension is the A-dimension, his Akashic dimension outside of space–time, which can be identified with Bohm's implicate, enfolded order.[349] According to László, there is continual two-way interaction between the A-dimension and the M-dimension. In a cybernetic cycle, the A-dimension provides the blueprint for manifesting changes occurring in the timespace M-dimension, while information recording these changes flows back to and are recorded in the nontemporal vibrations of the A-dimension.[350]

In this model of an A-field and M-field interacting can be seen the functional properties to support Rupert Sheldrake's morphic field paradigm. Sheldrake describes his theory of the morphic field's memory as morphic resonance, and here relates it to the quantum vacuum field:

> All morphic fields have an inherent memory given by morphic resonance. Morphogenetic fields, the organizing fields of morphogenesis, are one kind of the larger category of morphic fields, rather like a species within a genus . . . . Morphic fields must in some way interact directly or indirectly with electromagnetic and quantum fields . . . . Another possible point of connection between morphic fields and modern physics is through the quantum

[349] Bohm, *Wholeness and the Implicate Order*.

[350] László, *The Self-Actualizing Cosmos*, 95.

> vacuum field. According to standard quantum theory, all electrical and magnetic forces are mediated by virtual photons that appear from the quantum vacuum field and then disappear into it again.[351]

Sheldrake here states that some sort of memory information, stored in the morphic field outside of space–time, is interacting with electromagnetic fields within space–time.

László supports and extends Sheldrake's model to include a mechanism for morphic resonance, describing it as the effect of a long-term interaction of a species with the A-dimension.[352] Here is László's description of a species-specific pattern acting as a natural attractor for morphic resonance in the A-dimension.

> The information generated in this interaction is conserved in the A-dimension. The A-dimension is the memory of the M-dimension; it is the manifest world's Akashic record. The sea of Akashic information includes the species-specific pattern that is the natural "attractor" of healthy functioning in organism. This pattern results from the long-term interaction of a species with the A-dimension; it is the enduring memory of those interactions; and it codes the generic norms of viable species.[353]

[351] Sheldrake, *Morphic Resonance: The Nature of Formative Causation*, 158.

[352] László, *The Self-Actualizing Cosmos*, 57.

[353] Ibid.

László here parallels Sheldrake's description of morphic resonance, describing it as information transfer from a morphic field (A-field) to a living organism (M-field), a sort of information blueprint for the living organism:

> According to the Akasha paradigm the information that coordinates the functions of a living organism is a specific pattern in the sea of A-dimensional information. This corpus of information governs action, interaction, and reaction throughout the manifest world. It also governs the functions of the living organism. It is a blueprint of normal organic functioning.[354]

Sheldrake's morphogenetic field thus mirrors László's A-dimensional field, but while Sheldrake makes explicit a possible relationship or interaction of morphic fields with electromagnetic fields, László fails the attempt to identify the A-field with any of the current fields known to physicists. Here he states that the "Akashic Field" is, as yet, unknown to the physical sciences:

> The evidence for a field that would conserve and convey information is not direct; it must be reconstructed in reference to more immediately available evidence. Like other fields known to modern physics, such as the gravitational field, the electromagnetic field, the quantum fields, and the Higgs field, the in-formation field cannot be seen, heard, touched, tasted, or smelled . . . it seems evident that a further field is required. In

[354] Ibid., 56.

> my previous books I named the universal in-formation field the Akashic Field.[355]

László also puts forth in his theory that the A-field is the quantum vacuum itself, that the field is the fundamental energy and information-carrying field, interconnecting all things in and between all dimensions, nonlocally, in a multiverse. And László further describes the A-field, which he says is currently unknown to physics, as being a "vacuum-based holofield."[356] But László again declares, as did Bohm, that any such posited field must be firmly rooted in what science already knows about the nature of physical reality:

> The concept of such a field cannot be an ad hoc postulate, nor can it be an extra-scientific hypothesis. It must be rooted in what science already knows about the nature of physical reality . . . . We assume that the interaction of such a field with quanta and quanta-based systems—atoms, molecules, cells, organisms, ecologies, and systems even of cosmological dimension—produces nonlocal interaction within and among them.[357]

Unfortunately, while László states clearly that his concept must be rooted in science, he provides no clear connection between the A-field and the tenets of

[355] László, *Science and the Akashic Field*, 73.

[356] László, *Science and the Re-Enchantment of the Cosmos*, 34.

[357] László, *The Self-Actualizing Cosmos*, 13–15.

modern science, neither through mathematics nor physics, and this leaves his theory incomplete.

Nevertheless, László's intuited metaphor of the A-field being some sort of "blueprint" can be used to explain why evolution is not a random process. If we assume that through some mechanism the arc of the creating cosmos is guided by information stored in the A-field, then all of the "fine-tuned constants" discovered in our universe, and marveled at by cosmologists, might be more readily explained. If we assume that the A-field contains (and shares) information from billions upon billions of experimental attempts, in previously evolving space–time universes, to create new forms (new forms of particles, new forms of galaxies, new forms of matter, life, and consciousness), then it is the influx of this information collapsing from the A-field into the M-field that produces the exquisitely-tuned constants that are enjoyed in the structure of our particular universe. Such a viewpoint accords well with the recent ideas of the theoretical physicist Lee Smolin, that "for cosmology to progress, physics must abandon the idea that laws are timeless and eternal and embrace instead the idea that even they evolve in real time."[358]

László's theory brilliantly points the way towards a truly new paradigm for material science and future research directions, but it is unfortunate that his use of the Vedic term akasha and the Theosophist term "Akashic record" serves only to undermine his theory among the scientific community. His occasional use of the terms "spiritual" and "yoga" in his essays would

---

[358] Smolin, *Time Reborn*, 123.

seem also to limit the dissemination of his work within a broader scientific community, as seen here:

> Our bodily senses do not register Akasha, but we can reach it through spiritual practice. The ancient Rishis reached it through a disciplined, spiritual way of life, and through yoga.[359]

The "Akashic record" is a term invented and widely used by Theosophists in the mid-nineteenth century, in particular by Alice Bailey, in discussing Patañjali's *Yoga Sutras*, "The ākāshic record is like an immense photographic film, registering all the desires and earth experiences of our planet."[360]

The concept and term "Akashic record" is also frequently found in the early twentieth-century writings of the Austrian philosopher Rudolph Steiner, who identifies the Akashic records to be the source of information obtained through his supersensory perception into the history of the planet.[361] Steiner taught a theory, which he claimed to practice himself, through which, by conscious development of latent organs of perception, we can go beyond the ordinary space–time sensory systems into various modes by which we can directly access cosmic memory:

> Everything which comes into being in time has its origin in the eternal. But the eternal is not accessible to sensory perception.

[359] László, *Science and the Akashic Field*, 76.

[360] Bailey, *The Yoga Sutras of Patañjali*, 78

[361] Steiner, *Cosmic Memory: Prehistory of Earth and Man.*

> Nevertheless, the ways to the perception of the eternal are open . . . . We can develop forces dormant in us so that we can recognize the eternal . . . . In gnosis and in theosophy it is called the "Akasha Chronicle." Only a faint conception of this chronicle can be given in our language. For our language corresponds to the world of the senses. . .The one who has acquired the ability to perceive in the spiritual world comes to know past events in their eternal character.[362]

More recently, the American clairvoyant Edgar Cayce (1877–1945), known as "the sleeping prophet," and writing in the mid-twentieth century, also used the term "akashic records" to describe the source of ideas with which he claimed to have connected during his deep trance and sleep states.[363]

László's theories find a great deal of support in these writings and traditions outside of the scientific field of inquiry, based as they are upon participatory, experiential data, which is not normally of interest in the material sciences. He is not averse to discussing such participatory experience of his own, and using it to extend his theory:

> When we enter an altered state of consciousness images, ideas, and intuitions flow into our consciousness that transcend the range of our sensory perceptions. These elements are part of the totality of the information in the cosmic matrix: The

[362] Ibid., 39.

[363] Todeschi, *Edgar Cayce on the Akashic Records.*

> Akasha. This information is in a distributed form, as in a hologram.[364]

The idea of information distributed as a hologram accords well with Karl Pribram's research-based speculation on holographic memory processes; and, in the following passage, László's insistence on the importance of the need for understanding the whole, rather than exclusively separate parts, reminds us of David Bohm's explicate "whole," connected everywhere by the implicate order:

> The crucial feature of the emerging view is space and time transcending correlation. Space and time do not *separate* things. They *connect* things . . . information is conserved and conveyed in nature at all scales of magnitude and in all domains.[365]

László refers to the Akasha field as a "fifth dimension," and points out that it was seen in ancient culture, much as Bohm views the implicate order, to be the foundational basis of the cosmos:

> The *rishis* (seers) of India viewed the deep dimension as the fifth and most fundamental element in the cosmos; they called it by the Sanskrit term *Akasha.*[366]

---

[364] László, *The Self-Actualizing Cosmos*, 95.

[365] László, *Science and the Reenchantment of the Cosmos*, 35.

[366] László, *The Immortal Mind*, 113.

László's A-field hypothesis indicates that a shift in the material science approach to consciousness is occurring, and a new paradigm may be emerging:

> a paradigm where information rather than matter is seen as the basic reality, and where space and time, and the entities that emerge and evolve in space and time, are manifestations of a deeper reality beyond space and time.[367]

[367] Ibid.